A Set of Examples of Global and Discrete Optimization

Applied Optimization

Volume 41

A Set of Examples of Global and Discrete Optimization

Applications of Bayesian Heuristic Approach

by

Jonas Mockus

Institute of Mathematics and Informatics,
Kaunas Technological University,
Lithuania

KLUWER ACADEMIC PUBLISHERS
DORDRECHT / BOSTON / LONDON

A C.I.P. Catalogue record for this book is available from the Library of Congress.

ISBN 0-7923-6359-0

Published by Kluwer Academic Publishers,
P.O. Box 17, 3300 AA Dordrecht, The Netherlands.

Sold and distributed in North, Central and South America
by Kluwer Academic Publishers,
101 Philip Drive, Norwell, MA 02061, U.S.A.

In all other countries, sold and distributed
by Kluwer Academic Publishers,
P.O. Box 322, 3300 AH Dordrecht, The Netherlands.

Printed on acid-free paper

Printed in the Netherlands.

Contents

Preface

This book shows how the Bayesian Approach (BA) improves well-known heuristics by randomizing and optimizing their parameters. That is the Bayesian Heuristic Approach (BHA).

The ten in-depth examples are designed to teach Operations Research using Internet. Each example is a simple representation of some important family of real-life problems.

The accompanying software can be run by remote Internet users. The supporting web-sites include software for Java, C++, and other languages.

A theoretical setting is described in which one can discuss a Bayesian adaptive choice of heuristics for discrete and global optimization problems. The techniques are evaluated in the spirit of the average rather than the worst case analysis. In this context, "heuristics" are understood to be an expert opinion defining how to solve a family of problems of discrete or global optimization. The term "Bayesian Heuristic Approach" means that one defines a set of heuristics and fixes some prior distribution on the results obtained. By applying BHA one is looking for the heuristic that reduces the average deviation from the global optimum.

The theoretical discussions serve as an introduction to examples that are the main part of the book. All the examples are interconnected. Different examples illustrate different points of the general subject. However, one can consider each example separately, too.

In part I, About the Bayesian Heuristic Approach, some well-known results are discussed to help understand what the Bayesian Heuristic Approach is about. First the general ideas of Bayesian related approaches are outlined. This is for readers who want to understand the place of BHA in the general optimization theory and applications. Then it is shown how BHA works by considering a simple example of the knapsack problem. This is for those who are interested in optimization algorithms and would like to skip the underlying theory. In Part II, Software for Global Optimization, different versions of software implementing the Bayesian and other approaches to global optimization are considered. A set of portable Fortran 77 programs, reflecting the early developments of global and local optimization software, is discussed. An interactive Turbo C version, designed for DOS environment, and an interactive C++ version, designed for Unix environment, are described. Three Java versions of global optimization software are presented. The Java software is run by standard Internet browsers.

In part III, Examples of Models, examples of global and discrete optimization are described. The choice of examples aims to help teaching. The teaching topics include: Operations Research, Theory of Games and Markets, Decision Theory, Utility Theory, Queuing Theory, Scheduling Theory, Stochastic Optimization, and Discrete Optimization.

Following mathematical models are considered:

- competition models with fixed resource prices, the "Nash equilibrium,"
- competition models with free resource prices, the "Walras equilibrium,"
- models of dynamic competition,
- bimatrix game models, the "Inspector problem,"
- differential game models, the 'Star Wars' problem,
- investment models, the "Portfolio problem,"
- auto regressive moving average models (ARMA), the problem of exchange rate prediction,
- queuing models for prediction and optimization, the "Call Center problem",
- discrete optimization models, the scheduling problem,
- sequential statistical decision models, the "Bride problem."

The first nine models are solved using a set of algorithms of global and stochastic optimization. The last model is an example of stochastic dynamic programming. All the models are formulated in simple terms to serve as "classroom" examples. However, each of these models can be considered as simple representations of important families of real-life problems. Therefore, the models and the solution algorithms may be of interest for application experts, too.

This book is to describe underlying ideas of algorithms and models in mathematical terms supplemented by illustrative examples and explanations. The information is presented in two forms: hard copy and web-sites.

An advantage of the hard copy is that one does not need a computer to read it. The disadvantage is the static presentation. A hard copy cannot be updated. Therefore, new results must wait until the next edition.

The aim of supporting web-sites is to supplement the algorithms and models by software systems for different computing environments, using different languages and operating systems. An advantage of web-sites is the dynamic presentation. Thus, the results can be easily updated and made accessible free of charge for a wide range of readers. An additional advantage is the possibility to run the software directly by Internet. That is convenient teaching the programming using Internet.

The author considers the hard copy as a static introduction to the dynamic web-sites. There are two of them; they differ by volume and scope.

http : //mockus.org/optimum

http : //optimum.mii.lt/˜jonas

The software presented on these sites is intended for "active" readers, who would like to apply the results to their own problems immediately.

Contact e-mail addresses:

jonas@optimum.mii.lt

jonas@mockus.org

audris@bell − labs.com

Finally, I want to express my thanks to colleagues who directly or indirectly contributed to this book. I further acknowledge the support of the publisher in producing the book.

J. Mockus

January 2000

I

ABOUT THE BAYESIAN APPROACH

Chapter 1

GENERAL IDEAS OF BAYESIAN HEURISTIC APPROACH

1. OUTLINE

The traditional numerical analysis considers optimization algorithms that guarantee some accuracy for all functions to be optimized. This includes the exact algorithms. That is the worst case analysis. To limit maximal errors one needs computational efforts that often increase exponentially with the size of the problem. This is the main disadvantage of the worst case analysis.

The alternative is the average case analysis. Here an average error is not limited but is made as small as possible. The average is taken over a set of functions to be optimized. The average case analysis is the Bayesian Approach (BA) (Diaconis, 1988; Mockus, 1989a).

There are several ways to apply BA in optimization. The first one, the Direct Bayesian Approach (DBA), is defined by fixing a prior distribution P on a set of functions $f(x)$ and by minimizing the Bayesian risk function (DeGroot, 1970; Mockus, 1989a). The risk function $R(x)$ is the expected deviation from the global minimum at a fixed point x. The distribution P is considered as a stochastic model of $f(x)$, $x \in R^m$, where $f(x)$ might be a deterministic or a stochastic function. In the Gaussian case, assuming (Mockus, 1989a) that the $(n+1)$th observation is the last one

$$R(x) = \frac{1}{\sqrt{2\pi} s_n(x)} \times$$

$$\int_{-\infty}^{+\infty} \min(c_n, z) e^{-\frac{1}{2}\left(\frac{y - m_n(x)}{s_n(x)}\right)^2} dz. \tag{1.1}$$

Here $c_n = \min_i z_i - \epsilon$, $z_i = f(x_i)$. $m_n(x)$ is a conditional expectation with respect to observed values z_i, $i = 1, ..., n$. $s_n^2(x)$ is a conditional variance, and $\epsilon > 0$ is a correction parameter. The minimum of the risk function $R(x)$ is obtained (Mockus, 1989a) at the point

$$x_{n+1} = \arg\max_x \frac{s_n(x)}{m_n(x) - c_n}. \tag{1.2}$$

The objective of DBA, used mainly in continuous cases, is to provide as small average error as possible while keeping convergence conditions.

Another way, the Bayesian Heuristic Approach (BHA), means fixing a prior distribution P on a set of auxiliary functions $f_K(x)$. These functions define the best values obtained by using K times some heuristic $h(x)$. It is assumed that heuristics $h(x)$ depend on some continuous parameters $x \in R^m$. The heuristic helps to optimize an original function $C(y)$ of variables $y \in R^n$, where $m < n$ (Mockus et al., 1997). As usual, components of y are discrete variables. Heuristics are based on expert opinions about the decision priorities. Now both DBA and BHA will be considered in detail.

2. DIRECT BAYESIAN APPROACH (DBA)

The Wiener process is a common (Kushner, 1964a; Saltenis, 1971; Torn and Zilinskas, 1989) stochastic model in the one-dimensional case $m = 1$.

Figure 1.1 shows the conditional expectation, the conditional standard, and the risk function with respect to available evaluations. The Wiener model implies continuity of almost all sample functions $f(x)$. The model assumes that increments $f(x_4) - f(x_3)$ and $f(x_2) - f(x_1)$, $x_1 < x_2 < x_3 < x_4$, are stochastically independent. Here $f(x)$ is Gaussian $(0, \sigma x)$ at any fixed $x > 0$. Note, that the Wiener process originally provided a mathematical model of a particle in the Brownian motion.

The Wiener model is extended to multidimensional case, too (Mockus, 1989a). However, simple approximate stochastic models are preferable if $m > 1$. Approximate models are designed by replacing the traditional Kolmogorov consistency conditions. These conditions require the inversion of matrices of nth order for computing the conditional expectation $m_n(x)$ and variance $s_n^2(x)$ [1].

Replacing the regular consistency conditions by:

- continuity of the risk function $R(x)$
- convergence of x_n to the global minimum
- simplicity of expressions of $m_n(x)$ and $s_n(x)$

[1] The favorable exceptions are the Markovian processes including the Wiener one. Extending the Wiener process to $m > 1$ the Markovian property disappears.

the following simple expression of $R(x)$ is obtained using the results of (Mockus, 1989a)

$$R(x) = \min_{1 \le i \le n} z_i - \min_{1 \le i \le n} \frac{\|x - x_i\|^2}{z_i - c_n}.$$

The aim of DBA is to reduce the expected deviation. In addition, DBA has some good asymptotic properties, too. It is shown (Mockus, 1989a) that

$$d^*/d_a = \left(\frac{f_a - f^* + \epsilon}{\epsilon}\right)^{1/2}, \quad n \to \infty.$$

Here d^* is a density of points x_i around the global optimum. d_a is an average density of x_i in the feasible area. f_a is an average value of $f(x)$ in this area. f^* is an average value of $f(x)$ around the global minimum. ϵ is the correction parameter in expression (1.1). That means that DBA provides convergence to the global minimum for any continuous $f(x)$ and a greater density of observations x_i around the global optimum, if n is large.

Note, that the correction parameter ϵ has a similar influence as the temperature in simulated annealing. However, that is a superficial similarity. The good asymptotic behavior is just some "by-product" of DBA. The reason is that Bayesian decisions are applied for small size samples where asymptotic properties are not noticeable. To choose the next point x_{n+1} by DBA, one minimizes the expected deviation $R(x)$ from the global optimum (see Figure 1). The minimization of $R(x)$ is a complicated auxiliary optimization problem. That means that DBA is useful for expensive, in terms of computing times, functions of a few ($m < 20$) continuous variables. This happens in wide variety of problems.

Some examples are described in (Mockus, 1989a). These include maximization of the yield of differential amplifiers, optimization of mechanical systems of a shock absorber, optimization of composite laminates, evaluation of parameters of immunological models and nonlinear time series, planning of extremal experiments on thermostable polymeric compositions. A large set of test problems in global optimization is considered in (Floudas et al., 1999).

3. BAYESIAN HEURISTIC APPROACH (BHA)

The Bayesian Heuristic Approach (BHA) is for optimization of simple objective functions with large variable numbers. That is the case in many discrete optimization problems. As usual, these problems are solved using heuristics based on an expert opinion. Heuristics often involve randomization[2] procedures that depend on some empirically defined parameters.

The examples of such parameters are:
- an initial temperature of the simulated annealing,
- probabilities of different randomization algorithms in their "mixture."

In these problems DBA is a convenient tool for optimization of the continuous parameters of various heuristic techniques and their mixtures. That is the Bayesian Heuristic Approach (BHA) (Mockus et al., 1997).

Applying DBA, the expert knowledge is involved by defining a prior distribution. In BHA one exploits expert knowledge "twice". First time one does that by defining the heuristics. Second time the expert knowledge is involved by defining a prior distribution, needed to optimize parameters of heuristics using DBA.

4. ILLUSTRATIVE EXAMPLES

The example of a knapsack problem is convenient to illustrate basic principles of BHA in the discrete optimization. Given a set of objects $j = 1, ..., n$ with values c_j and weights g_j, find the most valuable collection of a limited weight g

$$\max_y C(y), \ C(y) = \sum_{j=1}^{n} c_j y_j, \ \sum_{j=1}^{n} g_j y_j \leq g.$$

Here the objective function $C(y)$ depends on n Boolean variables $y = (y_1, ..., y_n)$, where $y_j = 1$, if an object j is in the collection, and $y_j = 0$, otherwise. The well known greedy heuristic $h_j = c_j/g_j$ is the specific value of an object j. The greedy heuristic algorithm: "take the greatest feasible h_j", is very fast but it may get stuck in some non optimal decision. One may force the heuristic algorithm out of there by taking a decision j with some probability $r_j = \rho_x(h_j)$[3]

Here DBA is applied to optimize parameters x of the auxiliary function $f_K(x)$. This function defines the best results obtained applying K times a randomized heuristic algorithm $\rho_x(h_j)$. That is the most expensive operation of BHA. Therefore, the parallel computing of $f_K(x)$ should be used, if possible[4]. Optimization of x adapts the heuristic al-

[2]The randomization means that a decision depends on some random variable included into an algorithm deliberately.

[3]Meaning that a randomization is applied. For example:
- generate an uniformly distributed random number $\xi \in [0, 1]$
- take a decision j if $\xi < r_j$, where $\rho_x(h_j)$ is an increasing function of h_j and $x = (x_1, ..., x_m)$ is a parameter vector.

[4]In this case the solution time is inversely proportional to the number of parallel processors.

gorithm $\rho_x(h_j)$ to a given problem. The parameter x may be useful to solve related problems (Mockus et al., 1997). That helps using the "on-line" optimization.

We illustrate the parameterization of $\rho_x(h_j)$ by using three randomization functions $r_i^l = h_i^l / \sum_j h_j^l$, $l = 0, 1, \infty$. Here the superscript $l = 0$ denotes the uniformly distributed component. $l = 1$ defines the linear component of randomization. The superscript ∞ denotes the pure heuristics with no randomization. That means $r_i^\infty = 1$, if $h_i = \max_j h_j$ and $r_i^\infty = 0$, otherwise. The parameter $x = (x_0, x_1, x_\infty)$ defines probabilities of using the randomizations $l = 0, 1, \infty$, correspondingly. That can be understand as some "lottery of algorithms" where x_l is a probability to "win" the algorithm l. The detail description of the knapsack example is in the next chapter.

5. HEURISTICS

The main objective of BHA is to improve any given heuristic by defining the best parameters of some heuristic and/or the best "mixtures" of different heuristics. Heuristic decision rules, mixed and adapted by BHA, often outperform[5] the best individual heuristics, as judged by examples. In addition, BHA provides almost sure convergence. However, the results of BHA depend on the quality of specific heuristics. The quality of heuristics reflects the expert knowledge. That means, BHA should be considered as a tool for enhancing the heuristics but not for replacing them.

Many well-known optimization algorithms, such as the Genetic Algorithms (GA) (Goldberg, 1989), GRASP (Mavridou et al., 1998), and the Tabu Search (TS) (Glover, 1994), may be considered as some general heuristics that can be improved using BHA. There are specific heuristics tailored to fit particular problems. For example, the Gupta heuristic was the best one while applying BHA to the flow-shop problem (Mockus et al., 1997).

The Genetic Algorithms (Goldberg, 1989) is an important "source" of interesting and useful stochastic search heuristics. It is well known (Androulakis and Venkatasubramanian, 1991) that results of the genetic algorithms depend on mutation and crossover parameters. The Bayesian Heuristic Approach could be used in optimizing those parameters.

In the GRASP system (Mavridou et al., 1998), a heuristic is repeated many times. During each iteration, a greedy randomized solution is constructed and the neighborhood around that solution is searched for

[5]In terms of average errors.

the local optimum. The "greedy" component adds one element at a time, until a solution is constructed. A possible application of BHA in GRASP is in optimizing a random selection of a candidate to be in the solution. BHA might be useful as a local component, too, by randomizing the local decisions and optimizing the corresponding parameters.

In the Tabu search, two important parameters may be optimized using BHA. Those are:

- best combinations of short and long term memory,
- best balances of intensification and diversification strategies.

The examples show that the Bayesian Heuristics Approach is useful applying many well-known stochastic and heuristic algorithms of discrete optimization. The convergence conditions, if needed, are provided by tuning the BHA (Mockus et al., 1997).

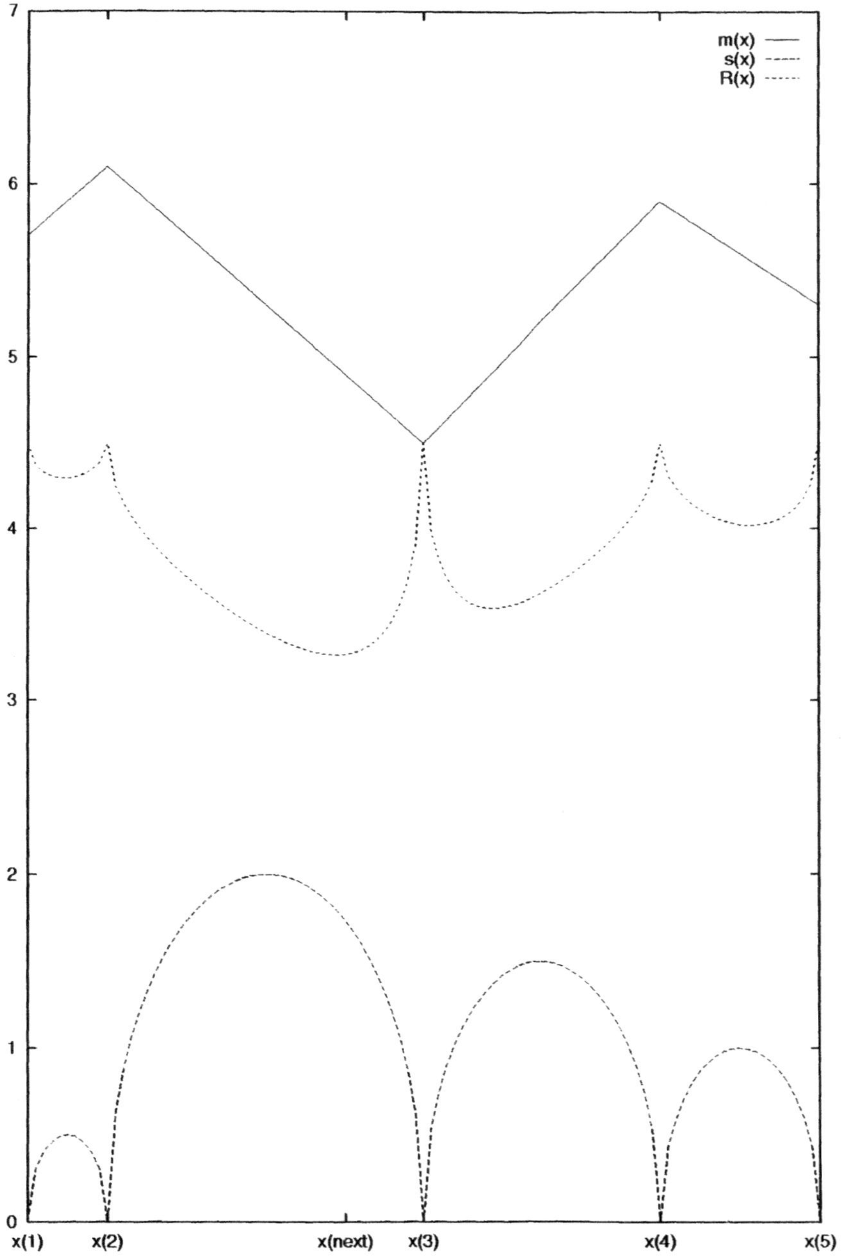

Figure 1.1. The Wiener model. The conditional expectation $m(x)$, the conditional standard $s(x)$, and the risk function $R(x)$ with respect to observed values $x(1), y(1), x(2), y(2), \ldots$

Chapter 2

EXPLAINING BAYESIAN HEURISTIC APPROACH BY EXAMPLE OF KNAPSACK PROBLEM

In the previous chapter we showed how BHA works while searching for the solution of the knapsack problem. Now we shall use this problem to illustrate similarities and differences of various approaches including BHA.

The reason for such special attention is that the knapsack problem is a simple example of important family of NP-complete problems[1]. Note, that this example is good mainly for illustration of BHA and alternative approaches. The reason is that the simple greedy heuristic of the knapsack problem is very efficient. Therefore, only marginal improvements can be made by more sophisticated methods, including BHA. In scheduling problems, such as the flow-shop and batch scheduling (Mockus et al., 1997), BHA works more efficiently. However, the scheduling problems are more difficult for explanations. Thus, we start the explanations by the knapsack problem.

The knapsack problem is to maximize the total value of a collection of objects when the total weight g of these objects is limited. Denote values of objects i by c_i and their weights by g_i.

$$\max_y \sum_{i=1}^{n} c_i y_i \tag{2.1}$$

$$\sum_{i=1}^{n} g_i y_i \leq g \tag{2.2}$$

$$y_i \in \{0, 1\}.$$

[1]No polynomial time algorithm has been found for solving the NP-complete problems. No proof is known that these problems cannot be solved in polynomial time.

Here the objective $C(y) = \sum_{i=1}^{n} c_i y_i$ depends on n Boolean variables $y_i,\ i = 1, ..., n$.

1. EXACT ALGORITHMS

1.1 EXHAUSTIVE SEARCH

The decision m means that a collection of objects $I(m) = \{i : y_i(m) = 1\}$ is selected. The value of the collection is $c(m) = C(y(m))$. This collection is described by the decision vector $y = y(m) = (y_i(m),\ i = 1, ..., n)$. The simplest exact algorithm is to compare all the collections:

1. select a current collection $I(m^N),\ N = 1, ...K$,
2. record the best current collection $I^*(m^N)$,
3. stop when all the collections are considered $N = K = 2^n$.

The best current collection $I^*(m^N)$ denotes the most valuable collection satisfying the weight limit that is obtained during N iterations. The best current collection is updated during the recording operation in step two. It is replaced by better current collection. The exhaustive search of all the decisions needs $K = 2^n$ of iterations. This means $T = C\ 2^n$ of time, where the constant C is the observation time (the CPU time needed to evaluate sum (2.1) and to test inequality (2.2)).

1.2 BRANCH & BOUND (B&B)

The efficiency of exact algorithms can be improved by the Branch & Bound (B&B) techniques:

1. define the decision tree,
2. define the upper limits of the branches,
3. cut the branch, if the upper limit is less then the best current collection,
4. stop, if there are no more branches to consider.

For example,

1. branch 0: $y_1 = 0$, branch 1: $y_1 = 1$,
2. the upper limit of branch 1:

$$C^1 = c_1 + \max_y \sum_{i=2}^{n} c_i y_i$$

$$g_1 + \sum_{i=2}^{n} g_i y_i \leq g$$

the upper limit of branch 0:

$$C^0 = \max_y \sum_{i=2}^{n} c_i y_i$$

$$\sum_{i=2}^{n} g_i y_i \leq g$$

3. cut branch 0, if $C^0 \leq c_1$,

4. stop, if there are no more branches to consider.

In this algorithm, the number of iterations to obtain the exact solution is $K \leq 2^n$ and the time is $T \leq C\ 2^n$. Usually, the time T is much less then the exact upper limit $C\ 2^n$. However, this limit may be approached. For example, if all the prices c_i and weights g_i are almost equal. In such a case no branches will be cut. That is a reason why approximate algorithms are preferred, if n is large.

2. APPROXIMATE ALGORITHMS

2.1 MONTE CARLO

The simplest approximate algorithm is Monte-Carlo (MC):

1. select a collection $I(m^N)$, $N = 1, ..., K$ with probability [2]

$$r(m) = \frac{1}{2^n}, \tag{2.3}$$

2. record the best current collection $I^*(m^N)$,

3. stop, if the number of iterations $N = K$.

Recording just the best current collection, one samples with replacement. Otherwise, one should keep in the memory up to 2^n of previous samples. In cases with replacement, one does not know when the optimum is reached. Therefore, MC stops, if the time allocated for the optimization is exhausted. This algorithm converges to an exact solution with probability one, if the number of iterations $K \to \infty$. However, the convergence is very slow, nearly logarithmic. Note, that the time $T = Cn^m$ is not a limit for MC with replacement.

[2]That can be understood as a lottery with the probability $r(m)$ to win a collection m. This lottery can be carried out by generating a random number ξ uniformly distributed in $[0, 1]$ and by selecting a collection m if $\xi < r(m)$.

2.2 GREEDY HEURISTICS

Define the heuristic h_i as the specific value of the object i

$$h_i = \frac{c_i}{g_i}. \tag{2.4}$$

This "greedy" heuristic is the well known and widely used in the knapsack problem. The Greedy Heuristic Algorithms prefer a feasible[3] object with the highest heuristic h_i:

1. select the object i with the best heuristic h_i in the starting list[4],
2. move the object from the starting list into the current one,
3. test feasibility of the collection which includes this object,
4. if the collection is infeasible, delete the object from the starting list,
5. test, whether the starting list is empty,
6. if the staring list is not empty, go to the step one,
7. if the starting list is empty, stop and record the current list as the best collection $I^*(m^N)$, where $N \leq n$ is number of the last iteration.

The greedy heuristic algorithm is fast, $T < C\ n^2$. The inequality is there because objects are gradually removed from the starting list. The algorithm may stick in some non optimal decision. The greedy heuristic h_i is obviously useless, if the specific values are equal $h_i = h,\ i = 1, ..., n$. In such cases the outcome depends on the numeration of objects.

2.3 RANDOMIZED HEURISTICS

LINEAR RANDOMIZATION

Heuristic algorithms are shaked out of non optimal decisions by choosing an object i with some probability r_i that is proportional to the specific value h_i,

$$r_i = \frac{h_i}{\sum_{j=1}^{n} h_j}. \tag{2.5}$$

This algorithm is similar to that of Greedy Heuristic:

1. select an object i from the starting list at random with the probability r_i,

[3] Satisfying inequality (2.2).

[4] Initially the starting list includes all the objects

2. move the object from the starting list into the current one,
3. test feasibility of the collection including this object,
4. if the collection is infeasible, delete the object from the starting list,
5. test, if the starting list is empty,
6. if the staring list is not empty, go to step 1,
7. if the starting list is empty, stop and record the current list as a current collection $I(m^N),\ N = 1, ..., K$,
8. record the best current collection $I^*(m^N)$,
9. start new iteration $N + 1$ by restoring starting list to the initial state (as a set of all the objects) and return to step 1,
10. stop after K iterations and record the best current collection $I^*(m^K)$.

This algorithm is better than the greedy one, in the sense that it converges with probability one when $K \to \infty$. The algorithm is expected to be better than Monte-Carlo, too, since it includes an expert knowledge by relating the decision probabilities to heuristics.

MIXED RANDOMIZATION

We may choose the preferred heuristic algorithm by considering all of them separately. Here the quality estimate of each algorithm is the best result obtained after K iterations. That is a traditional way.

One solves the same problem in a more general setup by considering a "mixture" of different heuristic algorithms. For example,

- denote the Monte Carlo by the index $l = 0$,
- denote the Linear Randomization by the index $l = 1$,
- denote the Greedy Heuristics by the index $l = 2$.

Then the Mixed Randomization uses the algorithm denoted by the index l with some probability $x(l)$ that is defined by probability distribution $x = (x(l),\ l = 0, 1, 2)$:

1. select an algorithm l at random with probability $x(l)$,
2. record the best current collection $I^*(m^N),\ N = 1, ..., K$,
3. start the new iteration $N + 1$ by returning to step 1,

4. stop after K iterations and record the best current collection $I^*(m^K)$, denote its value by $f_K(x) = c(m^K)$.

Here $f_K(x)$ is the best result obtained using K times the Mixed Randomization with the probability distribution x[5].

That is a way to extend a family of possible decisions from the discrete set $x(l) \in \{0,1\}, \ l = 0,1,2$ to the continuous one. That is important because the best results we often obtain using a mixture of algorithms (Mockus et al., 1997).

Here we mixed three approximate algorithms: the Monte Carlo, the Linear Randomization and the Greedy Heuristics. Following expressions define a larger family of different approximate algorithms, including these three

$$r_i^l = \frac{h_i^l}{\sum_{j=1}^n h_j^l}, \ l = 0,1,2,...,L, \tag{2.6}$$

and

$$r_i^\infty(m) \ = \ \begin{cases} 1, & \text{if } h_i = \max_j h_j, \\ 0, & \text{otherwise.} \end{cases} \tag{2.7}$$

Here the superscript $l = 0$ denotes the Monte-Carlo randomization. The superscripts $l = 1, \ l = 2, ..., l = L$ define a family of "polynomial" randomizations. The superscript ∞ denotes the greedy heuristics with no randomization. The Mixed Randomization means using randomization functions $r_i^l, \ l = 0,1,2,...,L,\infty$ with probabilities $x(l)$ and recording the best result $f_K(x)$ obtained during K iterations.

OPTIMIZATION OF MIXTURE

The optimization of the mixture x is difficult because $f_K(x)$ is a stochastic function and the multimodal one, as usual. A natural way to consider such problems is by regarding functions $f_K(x)$ as samples of a stochastic function defined by some prior distribution. The next observation is obtained by minimizing the expected deviation from the exact solution. This technique is called the Bayesian Heuristic Approach (BHA) (Mockus et al., 1997).

2.4 PERMUTATION

If we build a solution from "scratch," then we may apply so-called greedy heuristics (Helman et al., 1993). Building from scratch is convenient, if no initial expert decision is known. Otherwise, building from

[5]The algorithm is implemented in C++ and Java in the files *fi.C* and *Knapsack.java* correspondingly (see section 4. for details).

scratch, one ignores the expert information included into that initial decision. As usual, the permutation algorithms are used to improve initial decisions.

In the knapsack problem, the complete collection I is a set of all n objects. The initial collection is a set of objects $I(m^0) = \{i : y_i(m^0) = 1\} \in I$ satisfying the inequality $\sum_{i \in I(m^0)} g_i y_i(m^0) \leq g$. The value of this collection is $c(m^0)$. A complement of the initial set $I(m^0)$ is denoted $I^0(m^0)$. We denote permuted collections as $I(m^j)$, $j = 1, ..., J$ and values of these collections as $c(m^j)$. We define the direct permutation heuristics as

$$h_j = h(m^j) = c(m^j) - c(m^0). \tag{2.8}$$

To avoid confusion, the longer symbols $h(m^j)$ and $hn(m^j)$ are used sometimes instead of the short ones h_j and hn_j. We define the normalized permutation heuristics as

$$hn_j = hn(m^j) = \frac{h_j - A_j}{A^j - A_j}, \tag{2.9}$$

where $A_j = \min_{k=1}^J h_k$ and $A^j = \max_{k=1}^J h_k$. Normalized heuristics (2.9) can be easily converted into probabilities by expressions such as (2.6). Direct heuristics (2.8) are convenient in some well-known algorithms, such as Simulated Annealing. Here is an example of permutation algorithm using linear randomization:

1. make J feasible changes m_j^0, $j = 1, ..., J$ of the initial collection $I(m^0)$, by replacing randomly selected objects using this algorithm
   (a) select at random with equal probabilities an object j_1 from initial collection $I(m^0)$,
   (b) delete this object,
   (c) select at random with equal probabilities an object j_0 from complement $I^0(m^0) = \{i : y_j(m^0) = 0\}$ of the initial collection, skip this step if the complement is empty,
   (d) put the object j_0 into the initial collection,
   (e) test the weight limit,
   return to step one of the replacement algorithm, if the updated collection is to heavy,
2. define normalized permutation heuristics hn_j for all the changes $j = 1, ..., J$,
3. define probabilities r_j by expression (2.13),

4. select current collection $I(m^N)$, $N = 1, ..., K$ at random with probability r_j,

5. record the best current collection $I^*(m^N)$,

6. stop after K iterations and record the best current collection $I^*(m^K)$.

SIMULATED ANNEALING

The Simulated Annealing (SA) is a popular global optimization method. Features of SA, considering it as a permutation algorithm, are:
- only one permutation is made, $J = 1$,
- the direct heuristic is used

$$h_j = c(m^j) - c(m^0), \tag{2.10}$$

- if $h_j \geq 0$, the new collection $I(m^j)$is selected with probability one,
- if $h_j < 0$, the new collection is selected with probability r_j that declines exponentially,

$$r_j = \begin{cases} e^{\frac{h_j}{x/\ln(1+N)}}, & \text{if } h_j < 0, \\ 1, & \text{otherwise.} \end{cases} \tag{2.11}$$

Here N is the iteration number and x is the "initial temperature." The difference of this formulation from traditional simulating annealing algorithms is that we optimize the parameter x for some fixed number of iterations $N = K$. We disregard the asymptotic behavior. The asymptotic properties are beyond the scope of the Bayesian Heuristics Approach.

GENETIC ALGORITHMS

General Idea. Genetic Algorithms (GA) present a general and visual way of permutation and randomization. GA is a methodology for searching a solution space in the manner similar to the natural selection procedure in biological evolution (Holand, 1975). Each candidate solution is represented by a string of symbols. The set of solutions at some state m is called the population of the mth generation. The population evolves for a prescribed number M of generations. The basic structure processed by GA is the string. Strings are composed of a sequence of characters.

A simple GA consists of
- one reproductive plan, defined as the fitness proportional reproduction
- two genetic operators, defined as crossover and mutation.

The probability of selection is defined in proportion to the fitness of individuals

During the crossover operation one splits two selected strings at some random position s. Then two new strings are created, by exchanging all the characters up to the split point s. During the mutation operation one alters some string characters at random. A mutation is of l-th order, if one changes l elements during one mutation operation. Both the crossover and the mutation are feasible, if they satisfy all constraints.

It follows from this definition that GA may be considered as a special case of the Permutation and Randomization. Thus, one may include GA into the framework of Bayesian Heuristic Approach. We illustrate that by using the Knapsack problem as an example.

Explaining Genetic Algorithm by Knapsack Example. In the Knapsack example the string of the collection $I(m)$ is represented as the Boolean vector

$$y(m) = (y_i(m),\ i = 1, ..., n). \tag{2.12}$$

The fitness of feasible[6] collection m^j is $c(m^j)$. The probability to select a string m^j is

$$r_j = \frac{hn_j}{\sum_{k=1} hn_k}, \tag{2.13}$$

where hn_j is the normalized permutation heuristics

$$hn_j = hn(m^j) = \frac{c(m^j) - A_j}{A^j - A_j}. \tag{2.14}$$

Here $A_j = \min_{k=1}^J\ c(m^k)$ and $A^j = \max_{k=1}^J\ c(m^k)$.

During the crossover operation one splits two selected strings at some random position s, $1 < s < n$. Then two new strings are created, by exchanging all the characters up to the split point s.

During the mutation operation one produces J mutants by inverting randomly selected components y_j of the Boolean vector y. A mutation is of l-th order, if one changes l, $1 \leq l \leq n$ components during one mutation operation. A mutant is fertile, if it satisfies all the constraints.

Denote by $I(m^N)$ a current collection at Nth iteration. Denote by $I^*(m^N)$ the best current collection obtained during N iterations.

Here is an example of a simple GA algorithm written in BHA terms. The algorithm is similar to that of Permutation.

1. produce a number J of fertile mutants by replacing randomly selected objects using feasible changes m_j^0, $j = 1, ..., J$ of the initial collection $I(m^0)$,

---

[6]Assume for simplicity that all the non feasible collections are eliminated automatically.

2. define normalized permutation heuristics hn_j, $j = 1, ..., J$ for all these mutants using expression (2.14),
3. define probabilities r_j by expression (2.13),
4. select two mutants $I_j(m^+)$ and $I_j(m^-)$ at random with probabilities r_j[7],
5. select a split point s at random with probability $1/n$
6. inverse the components $y_i(m^+)$, $i \leq s$ and $y_i(m^-)$, $i \leq s$ of these two mutants,
7. update normalized permutation heuristics hn_j reflecting cross-over results
8. update probabilities r_j using (2.13)
9. select a current collection at random with the probability r_j,
10. record the best current collection $I^*(m^1)$
11. go to step 4 with the probability $x(0)$
12. produce J fertile mutants by feasible changes of the current collection,
13. define normalized permutation heuristics hn_j, $j = 1, ..., J$,
14. define probabilities r_j by expression (2.13),
15. select a current collection at random with the probability r_j,
16. record the best current collection $I^*(m^2)$ and return to step 11,
17. stop after K iterations and record the best current collection $I^*(m^K)$.

It is tacitly assumed that some segments of strings define specific "traits." In such cases, one may "enforce" good traits by uniting "good" segments. One may expose bad traits by uniting "bad" segments, too. Then, reproduction plans tend to favor the good traits and to reject the bad ones.

Thus, the crossover operation "improves" the population, if the "traits" assumption is true. If not, then the crossover may be considered merely as a sort of mutation. That may help to jump the area dividing separate

[7] If by chance the same mutant $m = m^+ = m^-$ will be selected then repeat the selection procedure.

"local" minima. However, the similar jump may be accomplished by high order mutations, too.

As usual (Androulakis and Venkatasubramanian, 1991), mutations are considered as more "radical" operations as compared with crossovers. That is correct, if one changes many elements of the "genetic" sequence during one mutation. This happens, if the mutation order l is close to the string length n.

The results of some real life examples of network optimization (Mockus, 1967) and parameter grouping (Dzemyda and Senkiene, 1990) show that low order mutations, when merely a few elements of the decision vector $r(k)$ are changed, work better. In such cases, the mutation may be considered as a less radical operation because fewer components of the vector y are changed, as compared with the crossover operation.

We may improve efficiency of the simple GA algorithm by using the crossover rate $x(0)$ (see step eleven of the algorithm) as an optimization variable in the framework of BHA. This was applied solving the Batch Scheduling problems (Mockus et al., 1997). If the crossover rate is considered as a time function (Androulakis and Venkatasubramanian, 1991), the parameters of this function can be optimized using BHA.

3. SOFTWARE EXAMPLES OF KNAPSACK PROBLEM

There are two software versions: one in C++ and one is Java. In the C++ version, prices c_i and weights g_i are generated randomly. In the Java case, c_i, g_i reflects a real life example (Malisauskas, 1998). The optimal probability $x(\infty)$ of the greedy heuristic $h_i = c_i/g_i$ is near to one, but not exactly one. Some degree of randomization helps to escape non optimal decisions. The randomized procedures $x(l), r_i^{(l)}$ are defined by (2.6) and (2.7). Both the software versions consider just three components.

The $C{+}{+}$ version uses the Monte Carlo, the Linear and the Quadratic randomizations. Corresponding optimization parameters are
$x = (x(0), x(1), x(2))$. The Monte Carlo and Linear randomizations and the Pure Greedy heuristic are applied in the Java version. Corresponding optimization parameters are
$x = (x(0), x(1), x(\infty))$.

We are looking for such probability distribution x that provides the maximal $f_K(x)$ after K iterations. Bayesian methods (Mockus, 1989a; Mockus and Mockus, 1990; Mockus et al., 1997) are used for that.

3.1 C++

The aim of the GMC version is to estimate average error of BHA. A set of knapsack problems with random prices c_i and weights g_i is considered. The results of BHA and the exact deterministic B&B algorithms are in Table 2.1.

Average results were obtained by repeating the optimization procedures $K = 100$ times at fixed parameters c_i, g_i and probability distribution x. In Table 2.1 N_0 is the number of objects. K is the number of repetitions. K_B is the total number of observations by the Bayesian method. K_E is the number of nodes considered by the exact method. f_B is the best result of the Bayesian method. f_E is the exact optimum. $\delta_B\%$ is the mean percentage error of the Bayesian algorithm. $x_B(n), n = 0, 1, 2,$ are optimized probabilities of different randomizations obtained using a Bayesian method. If the deviation of some solution does not exceed 5%, we call this a 5% solution. One can expect that the 5% solution satisfies the applications, where the level of data uncertainty is not less than 5%.

Table 2.1. Comparison of the Bayesian method and the exact one.

$K_B = 100$, and $K = 1$							
N_O	K_E	f_B	f_E	$\delta_B\%$	$x_B(0)$	$x_B(1)$	$x_B(2)$
50	313	9.56057	9.62347	0.654	0.0114	0.0280	0.9605
100	526	13.0703	13.1241	0.411	0.0316	0.0412	0.9271
150	771	16.6301	16.6301	0.000	0.0150	0.1945	0.7904
200	875	37.4665	37.4859	0.050	0.0315	0.0530	0.9437
250	568	53.7781	53.9000	0.226	0.0091	0.0511	0.9397
300	1073	28.3144	28.6106	1.034	0.0113	0.0835	0.9050
350	1416	30.4016	31.7527	4.254	0.0064	0.0646	0.9288
400	2876	32.1632	33.3192	3.469	0.0202	0.0452	0.9344
450	1038	105.467	105.578	0.105	0.0101	0.0149	0.9748
500	2132	39.3583	42.1047	6.521	0.0078	0.1556	0.8365

Table 2.1 shows that we need to consider from 313 to 2876 nodes to obtain the exact solution. Only 100 observations are needed to obtain the 5% solution by the Bayesian method. The deviation exceeds 1% only for three cases in ten. The average deviation is 1.67%.

Assume that roughly the same computing time is necessary for one node and for one observation. Then the Bayesian 5% solution is about three times "cheaper" as compared to the exact one, if the number of objects is fifty. If this number is 400, then the Bayesian 5% solution is almost thirty times cheaper. Other examples, in particular ones applying

BHA to a family of scheduling problems (Mockus et al., 1997) show higher efficiency of BHA.

3.2 JAVA

Here w e optimize a "mixture" x of the Monte Carlo randomization, the linear randomization, and the pure greedy heuristic. The aim is to show how BHA works while solving a real life knapsack problem. The example illustrates how to apply the Java software system for global optimization called as GMJ1. Therefore, several figures are included. They illustrate the input and output of GMJ1 graphical interface.

DATA FILE

The data represents the weights, the values, the numbers, and the names of inventory items of the "Norveda" shop that sells "Hitachi" electrical tools. Table 2.2 shows a fragment of data file 'norveda.txt'. Here fields, separated by spaces, denote these parameters of objects:

Weight	Value	Number	Name
3.8	2830	1	CNF35U
10.5	4170	2	CNF65U
11.5	3850	1	CM12Y
1.8	1500	2	CE16
1.7	1500	2	CN16
17.0	2100	4	CC14
20.9	2890	2	J9312N
0.9	330	8	D6SH
1.7	1170	10	D10YA
1.3	630	5	D10VC

Table 2.2. A fragment of data file 'norveda.txt.'

- weights g_i (double),
- prices c_i (double),
- numbers (integer),
- names (string).

IMPLEMENTING TASK

The algorithm is implemented as a *Task* of the GMJ1 system. Figure 2.1 shows a fragment of the file 'Knapsack.java'

```
public class Knapsack implements Task     { private double
norm_point[];            private int    number_of_objects;
public  double total_weight=10;         private int
object_taken[];          private float  object_cost[];
private float  object_weight[];         private double
object_ratio[];          private Random rnd = new Random();
public  String data_url = new String("http://norveda.txt");
private String data_url_old = new String();
```

Figure 2.1. A fragment of the knapsack program.

RUNNING GMJ1

Figure 2.2 (top) shows the input page.

On *Property* fields,

$TotalWeight$ is the weight limit, from 10 to 10.000,

$URL\ of\ data\ file$ is the URL address.

This data file is on the same server as the applet. If the data is not correct, the corresponding field turns red. In the black-and-white figure 2.2, red is shown as the dark shadow. Therefore, the incorrect URL address is not legible.

On the *Dimension* fields,

Min is the minimal value of $x(l)$,

Max is the maximal value of $x(l)$,

$Default$ is the default value of $x(l)$.

The values $x(l)$ show the proportions of each method. The mixture of three methods of picking objects is considered: the Monte Carlo, the Linearly Randomized Heuristics and the pure Greedy Heuristics. Probabilities x_l of methods l, $l = 1, 2, ...$ are related to the proportions $x(l)$ this way $x_l = x(l)/\sum_j x(j)$. Figure 2.2 (bottom) shows the output page that opens when the computing is completed. Here *Iteration* means the iteration number where the best value was reached. $F(x)$ defines the best value, "Monte Carlo." "Randomized Heuristics," and "Greedy Heuristics" show the optimal part of each of the three methods in the mixture.

Figure 2.3 (top) shows how the best value of $f_K(x)$ changes subject to the iteration number.

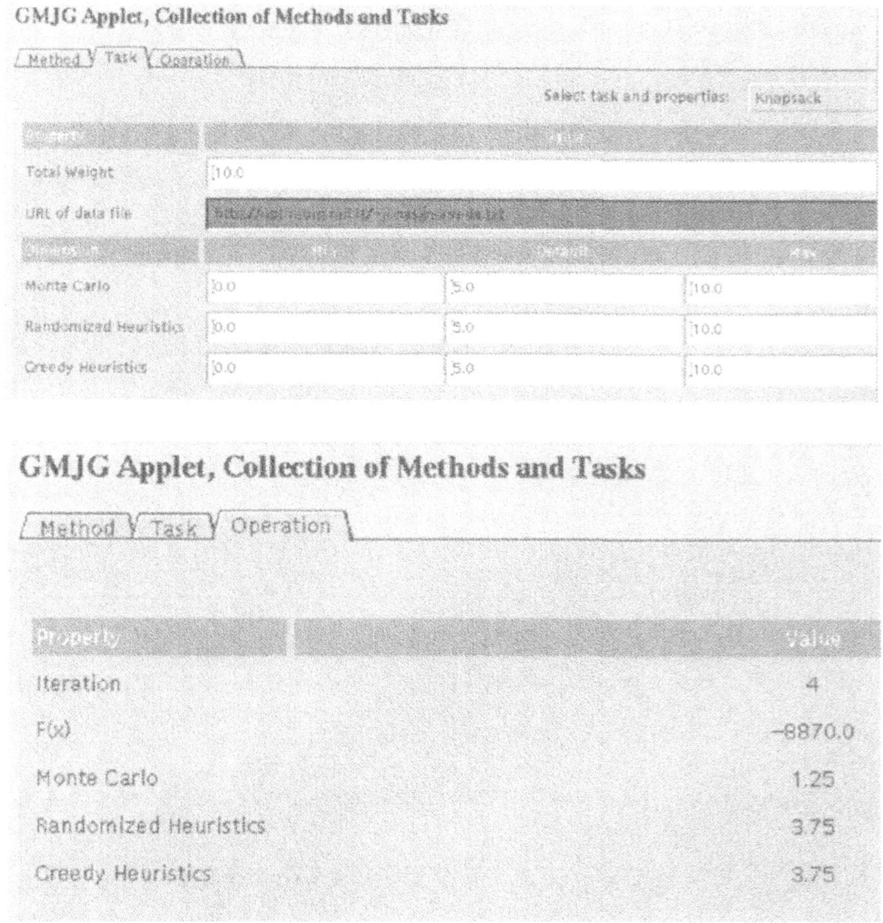

Figure 2.2. Input page (top) and output page (bottom).

Figure 2.3 (bottom) shows how $f_K(x)$ changes subject to proportions of the Monte Carlo randomization.

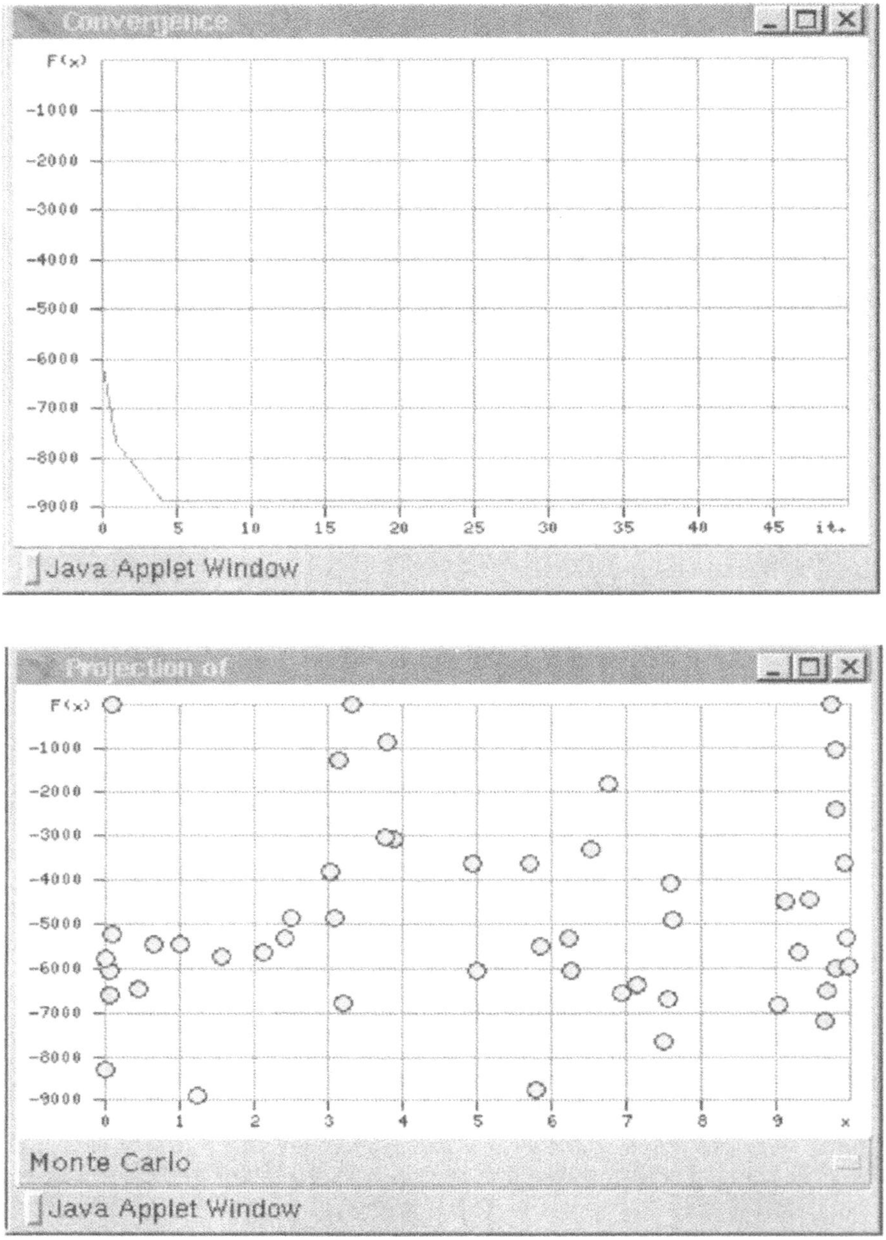

Figure 2.3. The best obtained value subject to iteration numbers (top), the objective function subject to proportions of the Monte Carlo randomization (bottom).

Figure 2.4 (top) shows how $f_K(x)$ changes subject to proportions of the linear randomization.

Figure 2.4 (bottom) shows how $f_K(x)$ changes subject to proportions of the pure greedy heuristics.

It is hard to notice any regularity in the *Projection* windows in Figures 2.3 (bottom), and 2.4. The reason is that all the variables change together during the optimization. To see good projections, one uses the method *Exkor*. Variables change one by one in this method (see Figures 8.4 and 8.5).

RESULTS

Table 2.5 shows how the results of optimization depend on the optimization method (the first column), the number of iterations (the second column), and the weight limit (the third column). The fourth column shows the iteration number where the best value was obtained, the fifth column shows the best value[8]. The sixth, the seventh, and the eighth column define the optimal mixture of three search methods. These mixtures can be expressed in percentage terms dividing each of the three numbers by their sum and multiplying by 100.

The results suggest that

- doubling the iteration number (from 500 to 1000) we lower the average error just by 3%,
- if the weight limit g is high, the greedy heuristic is almost useless (the optimal mixture $x = (7/15, 7/15, 1/15)$ meaning that the greedy heuristics part $x(\infty)$ is just $1/15$,
- if the weight limit g is low, the greedy heuristic is essential (the optimal mixture $x = (1/6, 1/6, 4/6)$ because the greedy heuristics part is ten times greater.

[8]The sign is minus, because by default GMJ reduces the objective of the Task object.

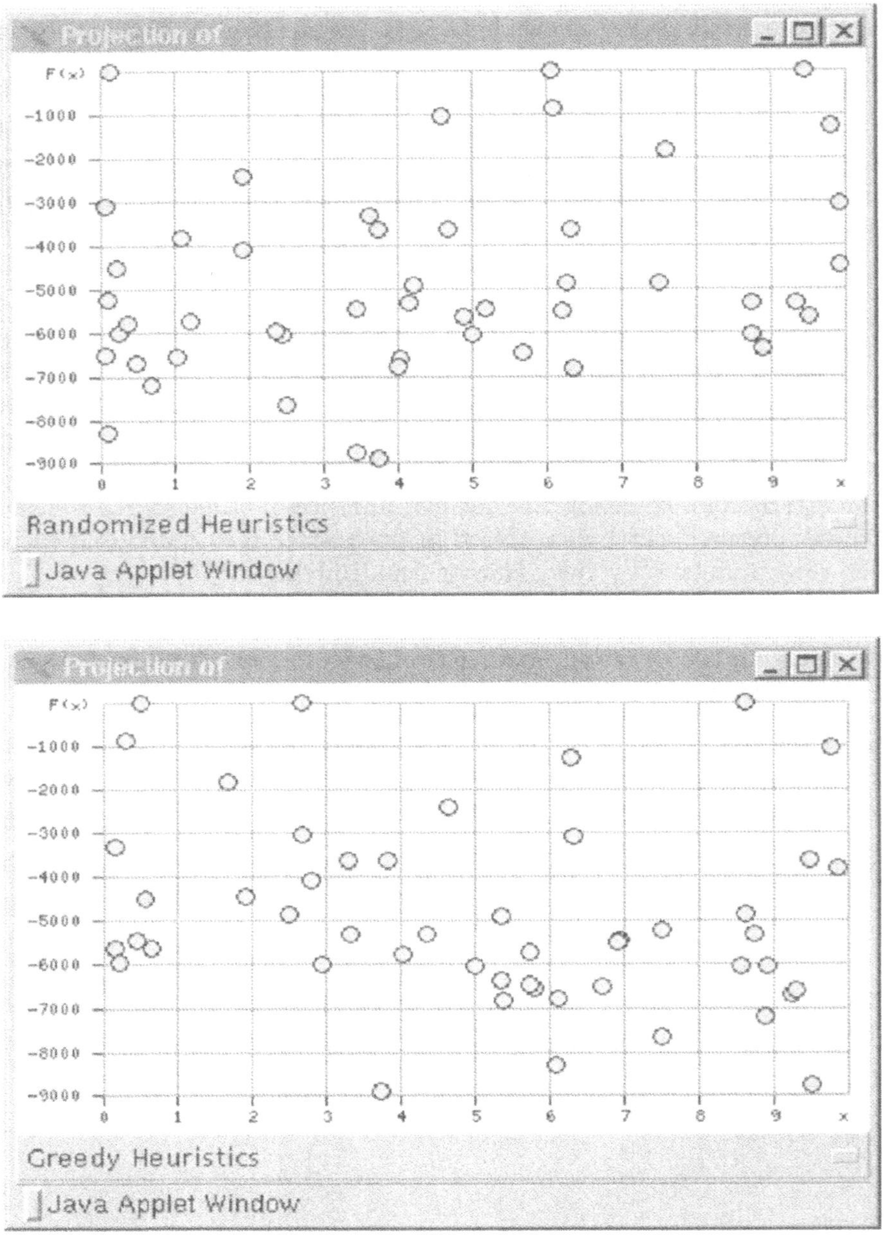

Figure 2.4. The objective function subject to proportions of the linear randomization (top), the objective function subject to proportions of the pure greedy heuristics (bottom).

Method	Iteration number	Maximal weight	Best iteration	Best function	Monte Carlo	Randomized Heuristics	Greedy Heuristics
Migl	1000	100	535	-66790	1,23	9,297	1,641
			854	-65920	2,578	9,548	8,733
			907	-65660	1,605	9,211	7,882
		1000	932	-488650	0,038	5,4	2,059
			735	-495230	0,047	3,296	0,445
			342	-488710	0,367	9,478	2,953
	500	100	152	-68940	1,237	9,533	6,913
			278	-66340	0,419	9,482	4,172
			269	-70360	4,334	9,528	5,683
		1000	483	-493640	0,103	7,129	1,504
			201	-487240	1,205	5,844	0,093
			439	-483170	2,542	6,819	0,634
Bayes	1000	100	851	-67640	2,731	8,574	9,539
			514	-65090	0,082	5,091	8,761
			595	-67490	1,265	9,508	2,819
		1000	741	-498850	0,473	9,947	0,644
			906	-504300	0,008	7,536	3,259
			949	-505170	0,073	8,612	0,198
	500	100	408	-65850	5,089	7,438	6,707
			213	-66890	0,018	8,213	5,339
			438	-66340	1,244	8,521	8,446
		1000	253	-495000	0,373	8,679	4,233
			37	-499860	0,417	5,745	0,239
			187	-502260	0,012	7,656	1,138

Figure 2.5. Table of results

II

SOFTWARE FOR GLOBAL OPTIMIZATION

Chapter 3

INTRODUCTION TO SOFTWARE

1. GENERAL DESCRIPTION

1.1 BACKGROUND

The software system Global Minimizer (GM) was initiated in early eighties. An initial set of algorithms was selected considering results of the international "competition" of different global optimization methods (Dixon and Szego, 1978). The experience in real life optimization problems was used updating the set. The set of global optimization algorithms includes

- four versions of the Bayesian search,
- a version of clustering,
- a version of the uniform deterministic grid,
- a version of the pure Monte Carlo search.

As usual, the global optimization is concuded by some local method. Exceptions are two global algorithms: the Torn version of clustering (Torn and Zilinskas, 1989), and the Zilinskas version of the Bayesian technique (Torn and Zilinskas, 1989). Both these algorithms contain some simple local search algorithms. The additional local search is not necessary for those two methods, but it may be useful.

There are three local optimization methods:

- a method of sequential quadratic programming for constrained optimization of smooth functions (Schittkowski, 1985),

- a simplex type method of Nelder and Mead with penalty functions for constrained optimization of non-differentiable functions (Himmelblau, 1972),
- a method of stochastic approximation type for "noisy" functions (Mockus, 1989a).

1.2 DESCRIPTION OF METHODS

GLOBAL METHODS

Bayes1 is the Bayesian method (Mockus et al., 1997).

Region : rectangular.

Objective : continuous (possibly with "noise").

Convergence : to global minimum (in probability, if noisy).

Comments :
for "expensive" objectives using not too many observations.

Mig1 is the Monte Carlo search.

Region : rectangular.

Objective : general.

Convergence : in probability.

Comments :
for inexpensive and irregular objectives using great number of observations.

Unt is the extrapolation type method (Torn and Zilinskas, 1989).

Region : rectangular.

Objective : continuous.

Convergence : to global minimum.

Comments :
for expensive objectives using not too many observations.

Exkor is the Bayesian coordinate line search method (Torn and Zilinskas, 1989).

Region : rectangular.

Objective : continuous.

Convergence : to global minimum on the search line.

Comments :
for approximately "separable" objectives and for preliminary exploration using the projection windows.

Glopt is the clustering method (Torn and Zilinskas, 1989).

Region : rectangular.

Objective : continuous.

Convergence : not provided.

Comments :
works well in many practical problems with a moderate number of local minima.

Lpmin is the uniform deterministic search (Sobolj, 1967; Dzemyda et al., 1984).

Region : rectangular.

Objective : general.

Convergence : to global minimum.

Comments :
for inexpensive objectives using many observations.

LOCAL METHODS

Nlp is the non linear programming method (Schittkowski, 1985).

Region : defined by linear and non-linear constraints.

Objective : differentiable.

Convergence : to local minimum.

Flexi is the simplex method (Himmelblau, 1972).

Region : defined by non-linear constraints.

Objective : non differentiable.

Convergence : not provided.

Lbayes is the stochastic approximation method with the Bayesian step size control (Mockus, 1989a).

Region : rectangular.

Objective : continuous with noise.

Convergence : to local minimum, in probability.

1.3 APPLICATION AREAS

Each subroutine represents a global or a local method. The choice of methods follows the comparability idea. That means that the computational complexity of the method should roughly correspond to that of the objective function:

- The Bayesian methods are recommended for "expensive"[1] functions. Those methods need auxiliary calculations to make each observation more efficient.

- For "cheap" functions, simple grid methods, like the Monte Carlo or the uniform deterministic grid (Sobolj, 1967), can be better. Here observations are not so efficient, but auxiliary calculations are negligible. This explains the relative efficiency of simple methods when optimizing simple functions.

- The clustering techniques (Torn and Zilinskas, 1989) may be the best choice, if we expect the number of local minima to be small.

- A simple Bayesian technique is available (Torn and Zilinskas, 1989) for global optimization of one-dimensional functions.

- There are optimization problems where objective functions can be roughly represented as a sum of components depending on different variables. Here the Bayesian method doing a line search along each coordinate shows good results, as usual. This method globally optimizes one variable at a time by the one-dimensional Bayesian search. Therefore, results depend on the starting point. This is the difference of this method from other methods of global optimization.

  However, a deviation from the global minimum can be made as small as desired by applying a multi-start search from different uniformly distributed starting points. The important advantage are good *projections*. They show how an objective function depends on different variables (see Figures 8.4 and 8.5). The other methods present messy projections (see Figures 2.3 (bottom), and 2.4). The reason is that all the variables change together.

1.4 CONSTRAINTS

All the global methods optimize in rectangular regions. Therefore, one represents linear and non-linear inequality constraints as penalty

---

[1] In a sense of computing time.

functions. The same applies to the local method of stochastic approximation type. In the local methods of simplex and sequential quadratic programming type, the linear and non-linear constraints are defined directly. This is done by constraint subroutines, supplied by users in addition to the objective function.

1.5 COMPUTING ENVIRONMENT

There are several versions of global optimization software

- a portable Fortran library,
- interactive software for Turbo C compilers and DOS operating systems,
- interactive software for C++ compilers and Unix operating systems,
- three versions of interactive software using Java (for JDK1.0, for JDK1.1, and for JDK1.2).

One may notice a cycle of portability in this sequence of software versions. The sequence is started by the portable Fortran library and is concluded by the Java systems. The systems between those two are more difficult to port. Fortran, Turbo C, and C++ versions are in (Mockus et al., 1997). All the software is on web-sites (see section 4. and Figure 3.1).

2. EARLY APPLICATIONS

Besides examples of this book, we briefly mention the examples considered in early publications (Mockus, 1989a; Mockus et al., 1997). Many of them are related to the optimization of parameters of mathematical models, represented as some systems of non-linear differential equations. The objective function $f(x)$ depends on a solution of these equations. Variables x represent the controlled parameters of the system. The following examples illustrate this family of problems.

- Maximization of general yield of differential amplifiers (Mockus, 1989a).
- Optimization of a mechanical system of shock-absorber (Mockus, 1989a).
- Evaluation of parameters of non-linear regression of an immunological model (Mockus, 1989a).

The last example suggests a broad area for applications of global optimization. It is well known that in non-linear regression the square

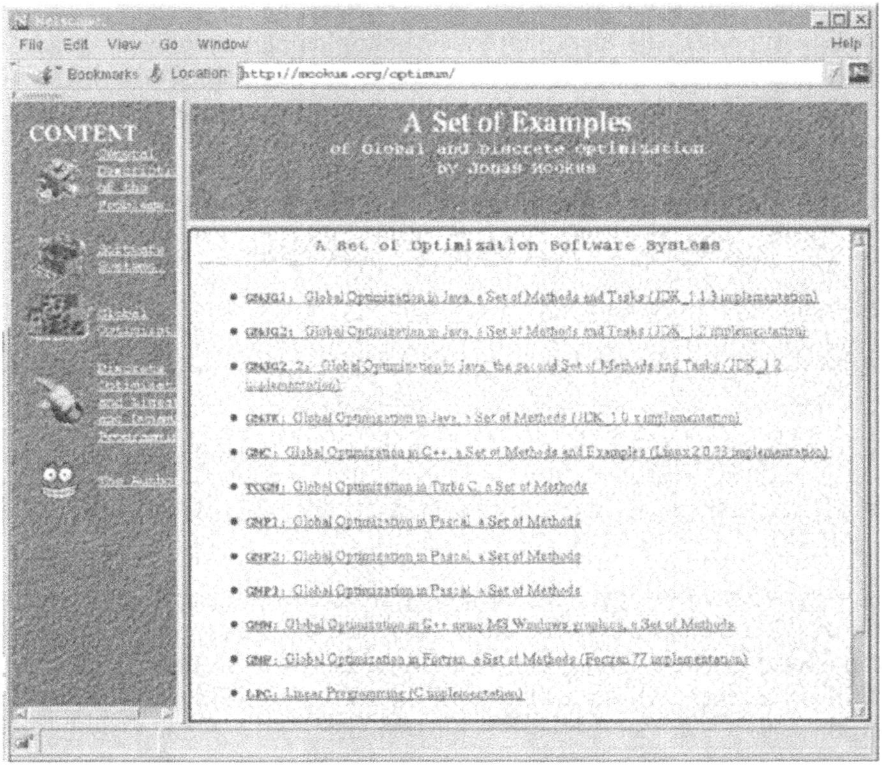

Figure 3.1. The Software Systems page.

deviation and the likelihood function could be multi-modal, for some data. The number of local minima may be very large, even in simple cases. An example is the evaluation of unknown parameters of ARMA models (Mockus et al., 1997). There are other important families of global optimization problems.

- The engineering design is a large source of difficult global optimization problems. One optimizes parameters of some mathematical models, non-linear, as usual. The optimization of composite laminates (Mockus et al., 1997) serves as an example.

- Many laws of nature could be defined in terms of global optimization. An example is the "Disk" problem: minimization of the potential energy of organic molecules (Mockus et al., 1997).

- Often one cannot describe the behavior of new materials and technologies by mathematical models, because the information and knowledge are not complete. Here one optimizes by changing control variables and observing the results of direct experimentation. An example is the planning of extremal experiments of thermostable polymeric compositions (Mockus et al., 1997).

The Bayesian algorithms of the continuous global optimization are an essential part of BHA. One applies them to optimize the parameters of randomized heuristics while solving various discrete optimization problems (Mockus et al., 1997).

3. DIFFERENT VERSIONS

3.1 FORTRAN LIBRARY, PORTABLE

The description of the global optimization software starts from the Fortran library (see Chapter 4). In other versions, mostly the same algorithms are used. Their parameters have a similar meaning. Therefore, the Turbo C, C++, and Java users may obtain some useful information by reading the Fortran library description.

The portable Fortran version can run on any computer with a standard Fortran compiler. Users represent objective functions as FUNCTION FI(X,N). Here X is an array of variables and N is its dimension. The lower bound array A and the upper bound array B define rectangular constraints. Using local methods of simplex and sequential quadratic programming type, one represents constraints by the subroutine
CONSTR(X,N,G,MC). Here, G is an array of length MC that contains values of constraints at a point X. MC is the number of constraints.

The advantage of the Fortran library is the portability. A disadvantage is limited interactive possibilities. Therefore, this version can be conveniently used for some well-defined optimization problems. For preliminary investigations of new problems, some interaction is essential.

3.2 TURBO C, INTREACTIVE

This version is for objective functions represented in C. It needs a Turbo C compiler. Users represent the objective function $f(x)$ as the C subroutine $fi.c$. Users can select a global or a local method by a menu system. Current results are observed in both, tabular and graphical, forms. There are two graphical forms.

GRAPH : shows how the best value of an objective function depends on observation numbers,

PROJECTION : shows how current objective function values depend on a variable.

The set of methods, in both the Fortran and the Turbo C versions, remains the same, except a non-linear programming method. The Biggs method (Biggs, 1975) is implemented in Fortran to solve non-linear programming problems. In all other languages, the sequential quadratic programming (Schittkowski, 1985) is used for these problems.

Good interactive facilities is an advantage of the Turbo C version. A disadvantage is that this version can be used only in the DOS-compatible environment. Often one prefers Unix systems while solving large scale optimization problems.

3.3 UNIX C++, INTERACTIVE

This is a global optimization software designed for UNIX and X - Window systems. It includes a usual set of interactive facilities. The main feature is the possibility to port to any environment that supports Unix and X-Window, including super computers and distributed computing systems.

3.4 JAVA JDK1.0, INTERACTIVE

This is a global optimization software in Java, designed for the JDK1.0 tool kit. The graphical interface is similar to that of the Unix C++ version. It can be used as a standalone application, in a way similar to that of the C++ version. The software can be run directly by Internet as an applet, too. Any Internet browser works, if it supports JDK1.0. That includes older versions of Netscape and Internet Explorer. This is an advantage, because these versions are still used.

3.5 JAVA JDK1.1, INTERACTIVE

This is a global optimization software in Java, designed for JDK1.1. The graphical interface is open and can be extended by including new methods, additional tasks and updated result analysis facilities. The software can be used as a standalone application or as an applet by the Internet browsers supporting JDK1.1. That excludes early browsers such as Netscape-3. Rectangular constraints are defined by the graphical interface. Other constraints are involved by adding the penalty functions that are defined by users.

3.6 JAVA JDK1.2, INTERACTIVE

This is a new version of global optimization software. It exploits new possibilities of JDK1.2. This Java version works on browsers that support JDK1.2. Otherwise, the JDK1.2 kit should be installed. Then the command

```
appletviewer http://mockus.org/optimum/gmjg2j/gmj.html
```

starts the applet *gmj.html* in the directory *optimum* of the server *mockus.org*. This way one bypasses a browser but must write the complete path.

The main new features:

- *constraint* functions can be involved by all optimization methods and other user defined functions,
- each objective function is accompanied by its own set of visualization algorithms and graphical interfaces, they appear only when this objective function is chosen.

The detail description of different software versions follows. Most versions use the same algorithms. Their parameters are similar. The software is on the web-sites.

4. WEB-SITES

Convenient reading at any place is an obvious advantage of hard copy. The theoretical considerations, including descriptions of mathematical models and algorithms, are clear and understandable, on a printed page. The static presentation is a disadvantage. A hard copy cannot be updated. Therefore, new results wait for the next edition.

The aim of the web-sites is to supplement the models and algorithms by interactive software systems. These systems can be tested directly by users in different computing environments, using different languages and operating systems (see Figures 3.1, 3.2, 3.3). The dynamic presentation is the important advantage of web-sites. Results can be easily updated and made accessible for many readers, free of charge. Therefore, the author considers this book as a static introduction to dynamic web-sites. There are two of them

1. *http : //mockus.org/optimum*
2. *http : //optimum.mii.lt/˜jonas*

The first web-site is designed for the international audience. Only selected items are included in this site. The updates are done by replacing the existing items, only when obviously better results appear. Therefore, updating is less frequent. The site is available always.

The second web-site is experimental. It is updated frequently, and has larger volume. The accompanying text, mostly in English, is in *txt*, *html*, *pdf*, or *ps* files. The site is used by graduate students of

Collection of Examples of Global Optimization

- Portfolio-GMJG-ARMA: Optimization of Investment in CD and Shares - example of global optimization by GMJ1 and ARMA(C)
- Portfolio-GMJE-GMC: Optimization of Investment in CD - example of global optimization comparing the performance of GMJ1 and GMC systems
- Portfolio-GMC: Optimization of Investment in CD- example of global optimization illustrating GMC system
- Walras-GMJ1: Competing Servers with Fixed Resources (Walras Problem) - an example of global optimization using GMJ1 system
- Duel-Java: Duel Problem -example of simple differential game using GMJ2
- Duel-Java: Duel Problem -example of simple differential game using Java (jdk1.1)
- Call Center: Call Rate Prediction Problem- an example of global optimization of time series model using ARMA-GMC system and call center data
- Shares: Stocks rate Prediction Problem- an example of global optimization of time series model using ARMA-GMC system and Lithuanian banks data
- ANN-GMC: Time Series Prediction Model - simple prediction model using ANN and GMC

Figure 3.2. Examples of Global Optimization.

Lithuanian universities. Therefore, a part of the text is both in English and in Lithuanian. Sometimes, during holidays this site is closed.

In the future, the web-sites will be regularly updated by new developments. However, the software will remain compatible with the present description.

Contact e-mail addresses:
jonas@optimum.mii.lt
jonas@mockus.org
audris@bell-labs.com

Collection of Examples of Discrete Optimization, Dynamic Programming, and Linear Programming

- Knapsack-GMJG: Knapsack Problem - an example of discrete optimization using heuristics and GMJG system
- Knapsack-GMC: Knapsack Problem - an example of discrete optimization using heuristics and GMC system
- Flow-shop-GMC: Flow-shop Problem - another example of optimal scheduling using heuristics and GMC system
- School -GMJG: School Scheduling Problem - an example of of optimal scheduling using heuristics and GMJG system
- School -GMJG: School Scheduling 2 Problem - another example of of optimal scheduling using heuristics and GMJG system
- School-1-GMC: School Scheduling Problem - an example of optimal scheduling using heuristics and GMC system
- School-2-GMC: School Scheduling Problem - an example of optimal scheduling using heuristics and GMC system
- Bride-1: Optimal Marriage Time Problem -first example of optimal sequential statistical decisions using dynamic programming implemented in Java, jdk1.1 version
- Buy-an-Appartment: Optimal Purchasing Time Problem -ant example of optimal sequential statistical decisions using the Bride-1 program
- Buy-a-Car: Optimal Purchasing Time Problem -ant example of optimal sequential statistical decisions
- Bride-2: Opt imal Marriage Time Problem -second example of optimal sequential statistical decisions using dynamic programming implemented

Figure 3.3. Examples of Discrete Optimization, Linear and Dynamic Programming.

Chapter 4

PORTABLE FORTRAN VERSION (GMF)

1. INTRODUCTION

The early Global Minimizer was in Fortran (GMF). Therefore, we describe the Fortran version as an introduction to later versions. The GMF system is to minimize a continuous function

$$f(x),\ x = (x_1, \dots, x_n) \tag{4.1}$$

where $x \in A \subset R^n$ (Mockus et al., 1997). It is assumed for most methods that the set A is a rectangular parallelepiped

$$A = \{z : a_i \leq x_i \leq b_i, \quad i = 2, \dots, n\}. \tag{4.2}$$

Other constraints are approximately reduced to (4.2) by the penalty function techniques.

A feature of all versions of the Global Minimum software, including GMF, is the Bayesian approach. The Bayesian methods are optimal, in the sense of an average deviation. Here the next observation is defined by minimizing the risk function $R(x)$. The risk is an expected deviation from the global minimum. The minimization of the risk function is justified, if the objective function $f(x)$ needs considerable computing. Otherwise, it could be better to do more observations, which need not to be planned optimally. In such a case, a good idea is to use a uniform search, for example, of LP type (Sobolj, 1967). Using the LP-search we lose some efficiency of observations but still have the convergence. However, the LP-search can be too expensive, if observations are very cheap. Then we apply the uniform random (Monte Carlo) search (Monte Carlo) which is very simple. Using the Monte Carlo we obtain the convergence only in probability. The LP-search (LPMIN) and the uniform random search (MIG1 and MIG2) algorithms. Both these algorithms are in GMF.

The method of clustering (Torn, 1990) is included as a good heuristic technique under the title GLOPT. It usually works well, if the number of local minima is small and known in advance, and if the "attraction area" of the global minimum is large. GLOPT is comparable to LPMIN by the computing time.

The global line search EXKOR is a "semi-global" method designed for approximately "additive" functions. EXKOR is convenient for visualization. One see clearly how values of the objective function depend on variables. That is important for preliminary investigations.

Global methods search the whole area, as usual. Otherwise, one can miss the global minimum. Obviously, the global search is not the most efficient way for local optimization. Therefore, in GM the global search is completed by the local one, as usual. The best result of a global search is defined by default as an initial point for a local search.

Methods of variable metrics type are widely used for local optimization of continuously differentiable functions without noise with non-linear constraints. The local mathod, called Recursive Quadratic Programming (Biggs, 1975), is included under the name REQP[1].

A specific Variable Metrics version for rectangular constraints (Tiesis, 1975) is called MIVAR4. For local optimization of non-differentiable functions with non-linear constraints the simplex type method (Himmelblau, 1972) is used under the name FLEXI.

Sometimes, one expects that the influence of some variables and their groups is considerably greater than that of others. Then the method LPMIN should be used for analysis of the structure before we start the regular optimization. The LPMIN orders variables by their "importance" using a sort of variance analysis techniques (Saltenis, 1989).

LBAYES is a stochastic approximation type algorithm with the "Bayesian" step length control. It is designed for local optimization of uni-modal functions with a noise.

In GMF, no machine-dependent routines are used. Therefore, the software can be adapted to any computer with a standard FORTRAN compiler by adjusting standard machine-dependent constants.

[1]In later versions a different algorithm (Schittkowski, 1985) is implemented under the name NLP.

2. GENERAL DISCUSSION

2.1 PARAMETERS

The parameters X, A, B, N, FM, IPAR, PAR, IPA, IPAA are used in all subroutines. Here
X is an array of length N which defines the coordinates of a point being considered (initial, optimal or current),
A is an array of length N which defines the lower bounds of X,
B is an array of length N which defines the upper bounds of X,
N is the number of variables (dimension of X) usually $N \leq 20$,
FM is the value of the function FI(X, N) at the point X,
IPAR is the array of length 30 which defines the real control parameters,
IPA is a shift of integer control parameters,
IPAA is a shift of real control parameters. If only one method is used, then both shifts are zero: IPA $= 0$ and IPAA $= 0$. If several methods are used sequentially, then a shift for the next method must be equal to the sum of numbers of control parameters used before by other methods. The number of control parameters of different methods is given in Table 4.1. For all the methods the first integer control parameter is the printing parameter: IPAR(IPA+1)=IPR
(IPA=0 if only one method is used). If IPR < 0,
then only diagnostic messages are printed.
If IPR $= 0$,
then the initial data, the final results and diagnostic messages are printed.
If IPR > 0,
then not only those but also the results of each IPR-th iteration are printed. The meaning of other control parameters will be explained when describing the corresponding subroutines.

2.2 LIST OF METHODS

Table 4.1 shows parmeters to combine methods into different sequences Table 4.1 shows that to set the algorithm parameters one performs

Table 4.1. Table of Parameters.

No	Name	Method	No. of control parameters Integer IPAR	Real PAR
		Global optimization with rectangular constants		
1	BAYES1	Bayesian	3	0
2	UNT	Extrapolation	4	0
3	LPMIN	Uniform deterministic	N+3	0
4	GLOPT	Clustering	3	0
5	MIG1, MIG2	Uniform random	2	0
6	EXTR	Bayesian one-dimensional	3	2
7	EXKOR	Extension of EXTR to multi-dimensional case	5	N+1
		Local optimization		
8	MIVAR4	Variable metric with rectangular constants	4	4
9	REQP	Variable metrics with non-linear constraints	4	4
10	FLEXI	simplex with non-linear constraints	4	2
		Local optimization with noise		
11	LBAYES	stochastic approximation with Bayes step length and rectangular constraints	3	2

many adjustments. This task is easy using "ready-made" examples. They include sequences of different algorithms. One selects needed algorithms just by "commenting-out" the redundant ones. For example, the file 'gmfl/ex8.f' in the enclosed disk includes seven algorithms named BAYES1, MIVAR4, LBAYES, FLEXI, REQP, UNT, GLOPT. The file 'gmfl/ex9.f' includes only one algorithm EXKOR. The software is on web-sites (see section 4.). Methods are described in (Mockus et al., 1997) and (Mockus, 1989a).

3. PROGRAM DESCRIPTION

3.1 COMMON BLOCKS

Values of the function FI(X, N) are in the array /BS1/Y(100). The array of variables X is defined by the array XN of length MN=N*M.
In the array /STATIS/IFAIL, IT, IM, M:
IFAIL is a control indicator:
- if the initial data is not correct, then IFAIL = 10 and return to the main program,
- if the initial data is correct, then IFAIL ≠ 10 and shows the number defining the stopping rule,
IT is the number of iterations,
IM is the number of the optional iterations,
M is the number of function evaluations (observations).

3.2 OBJECTIVE FUNCTIONS AND CONSTRAINTS

The function to be minimized should be represented as a real function FI(X,N). In most methods, only the lower and upper bounds are fixed by arrays A and B.
In methods with non-linear constraints, the subroutine
CONSTR (X, N, G, MC) is used. Here
G is a one-dimensional array of length MC that defines the constraints at the point X
MC is the number of constraints.

It is well known that local methods of optimization are sensitive to the scales of variables, as usual. The parameters of local methods in GMF are adjusted to the case when $A(I) = -1$ and $B(I) = 1,\ I = 1, N.$, as usual. Therefore, it is convenient to reduce the rectangular constraints $A \leq X \leq B$ to the N-dimensional hypercube $[-1, 1]^N$,. This can be done by the following reduction formulae[2]).

$$X(I) = \frac{2}{B(I) - A(I)} X0(I) - \frac{B(I) + A(I)}{B(I) - A(I)}, \ I = 1, N \ .$$

Here $X0$ are original variables, and X are variables scaled to fit into $[-1, 1]^N$.

Example. In most of the following examples, this test function $f(x)$[3] is used

$$f(x) = 2/N \sum_{i=1}^{N} (x_i^2 - \cos(18x_i)). \tag{4.3}$$

Here $N = 2$, $x_1 \in [-0.25;\ 0.5]$, $x_2 \in [-0.125;\ 0.625]$. The subroutine is FURASN.

```
      FUNCTION FURASN(X,N)
      DIMENSION X(N)
      NN=N
      F=0.
      DO 10 I=1,NN
      XI=X(I)
   10 F=F+XI*XI-COS(18.*XI)
      AN=NN
      FURASN=F*(2./AN)
      RETURN
      END
```

Figure 4.1. Objective Function FURASN

3.3 MAIN PROGRAM

The main program defines the input data and the sequence of methods. At the beginning, one selects (using Table 4.1 or the ready-made

[2]in LBAYES those formulae are included in the algorithm, therefore no additional reduction is needed.
[3]This is the well known Rastrigin test function (Rastrigin, 1968).

example 'gmfl/ex8.f'), a desirable sequence of methods of optimization and analysis. Then a function FI(X, N) is written that evaluates the objective function at a fixed point X.

If needed, the subroutine CONSTR(X, N, G, MC) is included. It evaluates the constraints at the point X. The length of arrays depends on the subroutines, as usual. The exception is the arrays IPAR and PAR. The length of these arrays is always thirty. The parameters of methods are included in the arrays IPAR and PAR by the given sequence of methods. Formal parameters IPA and IPAA are fixed using the rules that are given in the previous section. In the case, when only one method is used, IPA = IPAA = 0.

3.4 EXAMPLE OF THE MAIN PROGRAM

The global minimum of test function (4.3) is determined using the global Bayesian method BAYES1. Then the results of the global search are corrected by the local method of variable metrics MIVAR4.
The test function is represented as the function FURASN (X, N).
The arrays are: X, A, B, XN, HES, IPAR, PAR

It follows from Table 4.1 that, in the subroutine BAYES1, there are three integer parameters. In the subroutine MIVAR4, four integer parameters are used. In MIVAR4 four real parameters should be defined, too. This means that seven elements of IPAR and four elements of PAR should be fixed.
The main program is in Figure 4.2

3.5 INSTALLATION

MS DOS version of GMF. Installing:

- copy the file 'gmf.arj' (see section 4.).

- extract the GM files from the 'gmf.arj' archive

```
> arj e gmf.arj
```

```
      program main
      dimension x(2),a(2),b(2),xn(200),hes(3),
      ipar(30),par(30)
      data n,nm,nh,ipa,ipaa/2,200,3,0,0/
      data a/-0.25,-0.125/,b/0.5,0.625/
      data ipar/0,100,5,0,100,2,100,23*0/
      call bayes1(x,a,b,n,xn,nm,fm,ipar,ipa)
      ipa=3
      call mivar4(x,a,b,n,hes,nh,fm,ipar,par,ipa,ipaa)
      stop
      end
      function fi(x,n)
      dimension x(n)
      fi=furasn(x,n)
      return
      end
```

Figure 4.2. Example of MAIN

This version was tested using the Digital Research Fortran-77 Version 4.1 and LINK-86 Linkage Editor Version 1.5f, Digital Research, Inc. The example:

- compile

```
> f77 ex9.for
> f77 i1mach1.for
> f77 exkor.for
```

- link

```
> link86 ex9,i1mach1,exkor
```

- run

```
> ex9
```

Here the file 'ex9.for' is an example of the Main program using EXKOR method to minimize FURASN function (see expression 4.3).
The file 'exkor.for' is the source of EXKOR.
The file 'i1mach1.for' contains portability codes and other service routines, such as independent random number generator, various test functions, etc.

Linux version of GMF. Installing:

- copy the file 'gmfl.tgz' (see section 4.)
- extract the GM files from the 'gmfl.tgz' archive

  ```
  > tar -zxf gmfl.tgz
  ```

This version was tested using the standard Linux f77-style shell script 'f77' to compile and load Fortran and assembly codes. The example:

- compile and load

  ```
  > f77 ex9.f i1mach1.f exkor.f
  ```

- run

  ```
  > a.out
  ```

Note, that here the Fortran file extensions are 'f'.

Chapter 5

TURBO C VERSION (TCGM)

1. GENERAL DISCUSSION

A conceptual description of methods is given in (Mockus et al., 1997). Descriptions of real life examples of global optimization are in (Mockus et al., 1997) and in the next part of this book.

1.1 PURPOSE

The Turbo C version of the Global Minimizer TCGM is to solve global optimization problems similar to those of the Fortran version GMF. The optimization problem is defined in terms of variables describing the parameters that we can change, an objective function of those variables and some constraints restricting the possible change of variables. Formally, that is a nonlinear programming problem

$$\min_{a \le x \le b} f(x), \tag{5.1}$$
$$gi(x) \le 0, \ i = 1, ..., p$$
$$gi(x) = 0, \ i = p + 1, ..., m.$$

Here x is a vector of n variables. f(x) is an objective function. $a <= x <= b$ defines rectangular constraints. $gi(x) \le 0, \ i = 1, ..., p$ defines inequality constraints. $gi(x) = 0, \ i = p + 1, ..., m$ defines equality constraints. The objective function is deterministic or stochastic (for some methods). Rectangular constraints are defined for all the methods. The linear and non-linear constraints are defined just for two local methods: the Simplex by Nelder and Mead *Flexi* (Himmelblau, 1972), and the non-linear programming by Schittkowski *Nlp* (Schittkowski, 1985).

2. USER'S REFERENCE

2.1 REQUIREMENTS

- Turbo C version 2.0 or above.
- An experience of writing simple C functions.

2.2 INSTALLATION

- Copy the file *tcgm.arj* from a web-site (see Section 4.). Extract the GM files from the *tcgm.arj* archive

  ```
  > arj e tcgm.arj
  ```

2.3 INITIALIZATION

- Run the batch file GM.BAT using the command GM, and enter the Turbo C editor.
- Define the objective function as some C function with the name f. This function has two parameters:
  - double array of values of each variable,
  - the number of variables.
- Define the general linear or non-linear constraints as a C function named constr. It has four parameters:
  - double array of values of variables,
  - number of variables,
  - double array of values of constraints,
  - number of constraints.
- Compile file FI.C using directive Alt+C and correct possible errors.
- Link file FI.C to GLOBAL MINIMUM system using directive Alt+L.

- Exit Turbo C using directive Alt+X.

Now the GLOBAL MINIMUM is ready to minimize the problem that is defined in the file FI.C. Descriptions of objective functions and constraints have to be in the same file. Running GM.BAT, the file FI.C is displayed on the screen by Turbo C editor. Rectangular constraints are defined using the screen. Two algorithms, Nlp and Flexi, are exceptions. Here the constraints are defined by C function constr in the file FI.C. The four basic steps of initialization are

- run GM.BAT,
- edit file FI.C,
- compile FI.C,
- link file FI.C to system GLOBAL MINIMUM.

2.4 MINIMIZATION

The system will ask the following questions.

- Question: Number of variables.
  Action: Enter the number of variables of the problem.

- Question: Number of constraints.
  Action: Enter the number of linear and non-linear constraints of the problem (not counting the number of rectangular constraints that are defined interactively).

Now one see the GLOBAL MINIMUM copyright notice. GLOBAL MINIMUM is implemented as menu, dialog box and windows system. Each object can be either active or passive. There can be only one currently active object. A user can work only with active objects. There is a rectangular caption at the top of each object (menu, dialog box, or window). If this caption is bright, the object is active.

2.5 MENU SYSTEM

- Main Menu

- press any key and enter the MAIN MENU,
- select the type of optimization or exit the menu system selecting EXIT.

- Global Optimization Menu
  - select the GLOBAL item from the main menu,
  - select any of six available global optimization methods: LPMIN, MIG1, BAYES1, EXTR, UNT, GLOPT.

- Local Optimization Menu
  - select LOCAL item from the main menu,
  - select any of three available local optimization methods: NLP, FLEXI, or LBAYES. The initial point for local optimization is the result of previous global optimization, by default.

- Parameter Box
  - display of the PARAMETER DIALOG BOX follows the selection of method,
  - enter corresponding parameters of methods and press RETURN. Recommendations how to choose parameters are given in the description of methods (Mockus et al., 1997).

- Confirmation Menu
  - CONFIRMATION MENU appears following the selection of parameters,
  - RUN means run the method,
  - EXIT means return to LOCAL or GLOBAL OPTIMIZATION MENU.

- Operation Menu
  - OPERATION MENU appears following the selection of RUN item,
  - RUN means run the method, in case if it was stopped by selecting STOP item,
  - STOP means temporary stopping of the method and activating the OUTPUT DIALOG BOX,
  - press ENTER to return to the OPERATION MENU and to make the OUTPUT DIALOG BOX inactive,

- EXIT means exit the method and return to GLOBAL or LOCAL in the OPTIMIZATION MENU,
- ENTER means return to GLOBAL or LOCAL in the OPTIMIZATION MENU, after the method finishes optimization,
- RECORD means turn RECORD WINDOW on/off,
- PROJECTION means turn PROJECTION WINDOW on/off.

■ Information Display

- OUTPUT DIALOG BOX shows the results of each iteration (usually best point and goal function value at this point) when the method runs (see Figure 5.1),

Result					
Point	0.29	1.7	0.20	1.5	0.077
Value	0.26				

Figure 5.1. Table of TCGM results.

- RECORD WINDOW shows how the best function value depends on the iteration number (see Figure 5.2),

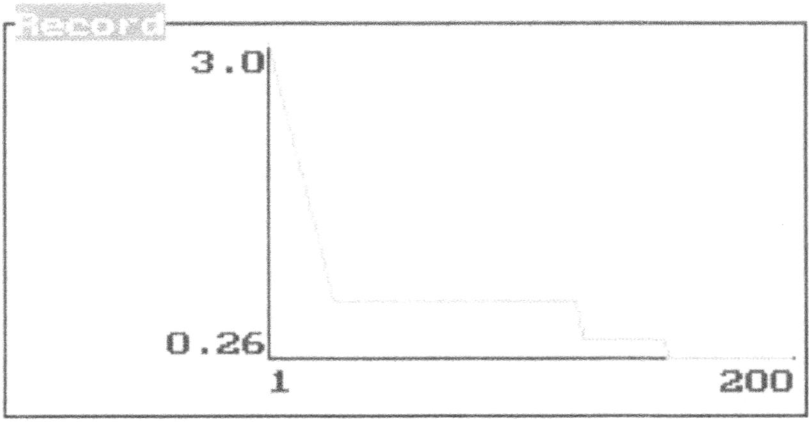

Figure 5.2. Graph of TCGM results.

- RECORD WINDOW is on when the method is started,
- if the graph on RECORD WINDOW is horizontal for a long time it can suggest that the minimum is reached or that the method

exhausted itself for this problem and it is better to try another method,

- if one selects RECORD item in OPERATION MENU while running the method then the RECORD WINDOW becomes invisible, if it was visible, and vice versa,
- PROJECTION WINDOW shows the projection of the observed (calculated) objective function values, the numbers of projection planes correspond to the numbers of variables,
- if one selects a PROJECTION item in OPERATION MENU while running the method, then one will see projection menu consisting of two items: ON and OFF, meaning on/off the PROJECTION WINDOW,
- if one selects ON then the projection dialog box appears and we can enter numbers defining the projection planes,
- select the number, press ENTER and one see how the observed values of the goal function depend on the corresponding values of the selected variable, the values of all other variables will not be seen.

- RESULT DIALOG BOX shows the results of calculations after the method finishes optimization.

2.6 NAVIGATION

- On Screen
  - screen is divided into three areas: bottom, middle and top, the functionality of these areas is different,
  - bottom area is for general information, such as description of variables, displaying error messages, displaying the state of an active object,
  - middle area is only for displaying and editing variables,
  - the top is a working area; menus, dialog boxes and windows are displayed here.
- On Menu
  - if some menu item is invisible, one opens it using corresponding cursor keys: UP, DOWN, LEFT, RIGHT,
  - the contents of the menu will scroll until the last available item will be reached.

- Dialog Box
  - on the DIALOG BOX one can see such variables as the parameters of method, coordinates of the best and current point, best and current values of goal function and so on, to view a variable one selects the variables name,
  - if the variable is a scalar, one see the corresponding value, next to the variable's name, the description of this variable is displayed at the bottom area of a screen, the exact value of the variable is displayed at the middle area,
  - if the variable is a vector then, after selecting the variables name, one can select the corresponding component of the vector using keys LEFT and RIGHT, the value of the selected component will be in the middle area, the number of component will be in the bottom area, along with the description of variable,
  - to see invisible components of the vector, use the keys LEFT and RIGHT, the vector scrolls, if there are invisible components,
  - press SPACE, to edit a variable, press ENTER, to exit the editor.

2.7 MOVING

- Press the F3 functional key and the red border will appear, move the border with keys UP, DOWN, RIGHT, LEFT and press ENTER to stop.
- The previous image of the object will be deleted and the object will be placed inside the new border.

2.8 SIZING

- Press the F4 functional key, move the object and press ENTER. The red border appears.
- Size the border using the keys UP, DOWN, RIGHT, LEFT and press ENTER to stop.
- The previous image of the object is deleted and the object is placed into new borders.

No changes happens, if borders are greater than maximal size of the object, or less than its minimal size.

Chapter 6

C++ VERSION (GMC)

1. PURPOSE OF GMC

The Global Minimizer's C++ version GMC is for continuous global optimization (Mockus et al., 1997). Users define their optimization problems describing

- parameters that can be changed (defined as variables x),
- function of those variables (defined as objective function $f(x)$),
- constraints restricting a possible change of variables (defined as inequalities $gi(x) \leq 0, i = 1, ..., p$, or equalities $gi(x) = 0, i = p + 1, ..., m$).

Formally

$$min f(x), \tag{6.1}$$

$$a \leq x \leq b \tag{6.2}$$

$$gi(x) \leq 0, i = 1, ..., p, \tag{6.3}$$

$$gi(x) = 0, i = p + 1, ..., m. \tag{6.4}$$

The objective function $f(x)$ may be deterministic or (for some methods) with "noise." As usual, the optimization region is defined by rectangular constraints (4.2). For some methods, Nlp and Flexi, this region is defined by linear and non-linear constraints (6.3) and (6.4).

2. USER'S REFERENCE

2.1 REQUIREMENTS

- C++ compiler, for example, GNU g++ compiler or any ANSI C++ compliant compiler,
- X-Window system,
- UNIX (Linux, Solaris, AIX),
- experience in writing simple C, or C++ functions.

2.2 INSTALLATION

Differences between various UNIX versions are reflected in the file 'Makefile'. An example for individual Linux users is in Figure 6.1. An AIX version for individual users is in (Mockus et al., 1997). A classroom (collective user) version for Solaris is there, too. Using some other UNIX version, the Makefile has to be changed accordingly. Installing GMC one should

- copy the archive file *gmc.tgz* from the web-site (see section 4.) by the unix command

  ```
  > mcopy a:  gmc.tgz
  ```

- extract GM files from the archive 'gmc.tgz' by the command

  ```
  > tar -zxf gmc.tgz
  ```

2.3 INITIALIZATION

- Define the objective as some function fi. The function fi has two parameters: an array of values of each variable and the number of variables. The function fi returns the value of objective, given the values of variables.
- Define "non-interactive" linear or non-linear constraints as a function named *constr* It has four parameters: an array of values of variables, the number of variables, an array of values of constraints, and the

number of constraints. Values of constraints are returned by the function *constr*.
A user can also define rectangular constraints interactively, during the optimization

- Describe objectives and constraints in a file named *fi.C* (see Figure 6.2)
  - the objective function is implemented by the routine

    ```
    double fi (const double *xx, int dim)
    ```

  - the constraints are defined by the routine

    ```
    void  constr (const double *x,  int n, double *g,
    int ng)
    ```

- Start optimize the problem described in the file *fi.C*.
- set X-window
- compile

  ```
  > rm  fi.o >  make
  ```

- run

  ```
  >  ./test
  ```

2.4 MENU SYSTEM

The opening menu of GMC is in Figure 6.3

- The Main Menu:
  Global, Local, Operations, Quit, Parameters, Results, Output.
- Global Optimization Menu:
  Bayes1, Migl, Unt, Exkor, Glopt, Lpmin.
- Local Optimization Menu:
  Nlp, Flexi, Lbayes.
  By default the initial point for local optimization is the result of previous global optimization.

- Operations Menu:
  Run, Stop, Exit.

- Quit Command.

- Parameter Box.
  Enter the parameters of methods and point O.K.
  An example of the parameter box is in Figure 6.3.

- Results Box:
  Shows the best objective Y and the best variables X(I). If no changes are needed, point O.K.

- Output Menu:

  - Convergence, Projection, and Numeric Windows.
  - Convergence window shows how the objective depends on the iteration number (see the Figure 6.4.
  - Projection windows show how the objective depends on the different variables. One defines the projection by entering the number of variable. Figures 6.5 and 6.6 show two projections.
  - Numeric window shows the current values of objective and variables (see Figure 6.7).

- Using Menu:

  - To see "invisible" components of a vector, point to 'down' or 'up' arrows. The vector scrolls, if there are invisible components.
  - To edit a variable, touch BACKSPACE[1], then edit the variable and point to O.K.
  - Enter the Parameter Box immediately after selecting a method.

[1] One should use the 'xmodmap' command to adapt the keymap, if the BACKSPACE key is not working.

```
#this is for linux 2.0.0
# C++ compiler. If error is "Can not load ..."
# (try CC or gcc instead of g++)
C++ = g++ -I$(INCDIR) -L$(LIBDIR) -O2
# pathname to X11/*.h files
# if compilation errors are "Can not find include file ..."
 then
# try /usr/local/include or /usr/include/X11R5
INCDIR      = /usr/include
# pathname to libX11.a
# If error is "Can not find library ..."
# try /usr/local/lib or /usr/lib/X11R5
LIBDIR      = /usr/X11/lib
CPP         = /lib/cpp
F77         = f77
LIBS        = -lX11 -lm
OBJS        = test.o WINDOW.o MENU_ITEM.o  SMENU_ITEM.o
 work.o MENU.o DBOX.o CONTROL.o STATIC.o EDIT.o BUTTON.o
 LBOX_ITEM.o LBOX.o     SCROLL_BAR.o SBBUTTON.o
 OUTPUT.o mig1.o MIG1.o exkor.o  task.o
 bayes1.o SUBMENU.o BAYES1.o SHIT.o
 CLIENT.o lbayes.o PAR_DBOX.o ERROR.o
 PROJ.o VAR_DBOX.o unt.o UNT.o EXKOR.o RES_DBOX.o
 VEDIT.o CONV.o fi.o NUM.o shit.o IO_DBOX.o LBAYES.o
  lpmin.o LPMIN.o glopt.o GLOPT.o flexi.o FLEXI.o HELP.o
RES         = menu.res bayes1p.res nlpp.res var.res
lbayesp.res
     res.res num.res mig1p.res untp.res exkorp.res
 lpminp.res
             gloptp.res flexip.res help.res
TARGET      = test
.SUFFIXES: .C .o .rc .res .f
.f.o:   $(F77) -c $*.f
.C.o:
$(C++) -c $*.C
.rc.res:
$(CPP) -P $*.rc > $*.res
all: $(TARGET) $(RES)
$(TARGET): $(OBJS) Makefile
$(PURIFY) $(C++) -o $(TARGET) $(OBJS) $(LIBS)
# DO NOT DELETE THIS LINE -- make depend depends on it.
```

Figure 6.1. Example of 'Makefile' for Linux.

```
/* File: fi.C */
#include "fi.h"
#include <math.h>
int number_of_variables =2;
/* Objective function*/

double fi (const double *x, int n)
{
   int    i;
   double y = 0;

   for (i = 0; i < n; i++) {
double z = x [i];
y += z * z - cos (18. * z);
}
return 2 * y / n;
}
/*Constraints */

void constr (const double *x, int n, double *g, int ng)
{
/*
ng > 0 - calculates all constraints
ng < 0  - calculates only constraint with number (-ng)
(in this illustrative example there are no constraints,
except interactively defined ones)
*/
}
```

Figure 6.2. Example of the file 'fi.C.'

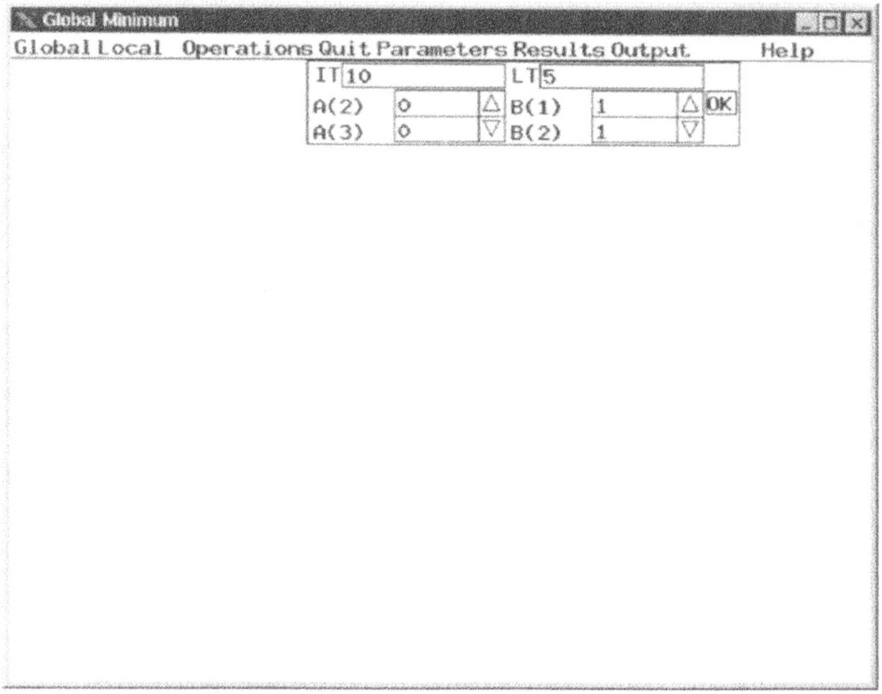

Figure 6.3. **An example of GMC opening menu and parameter box.**

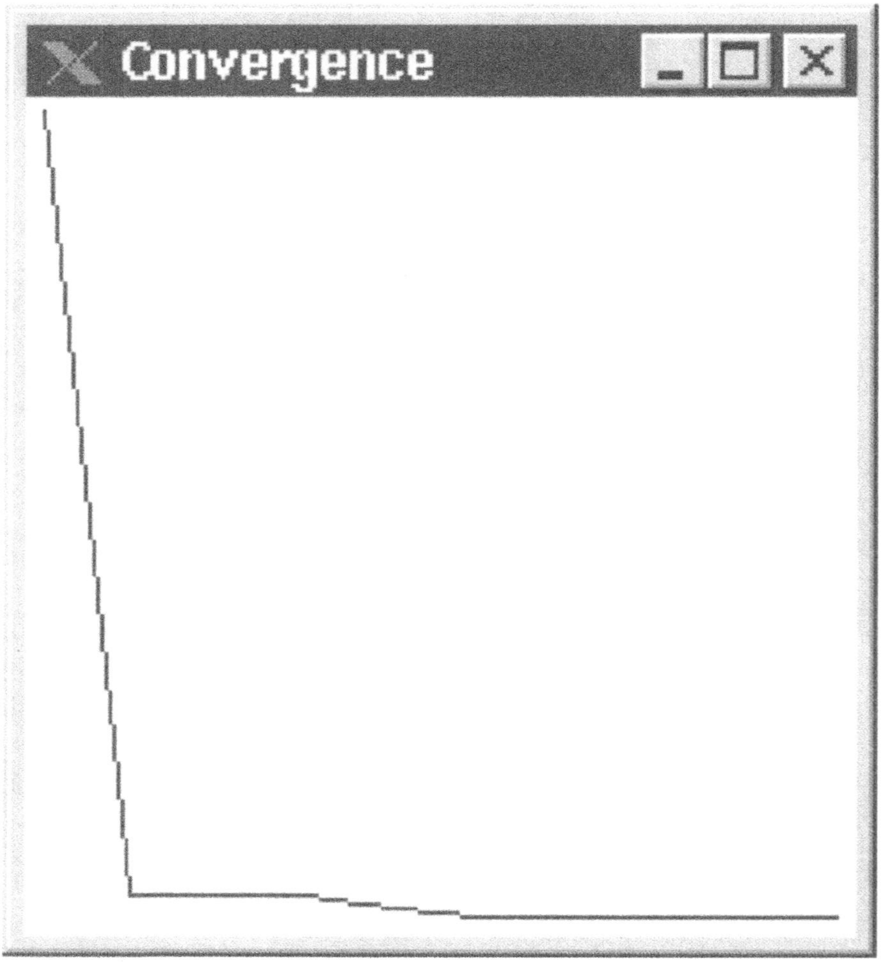

Figure 6.4. Convergence window.

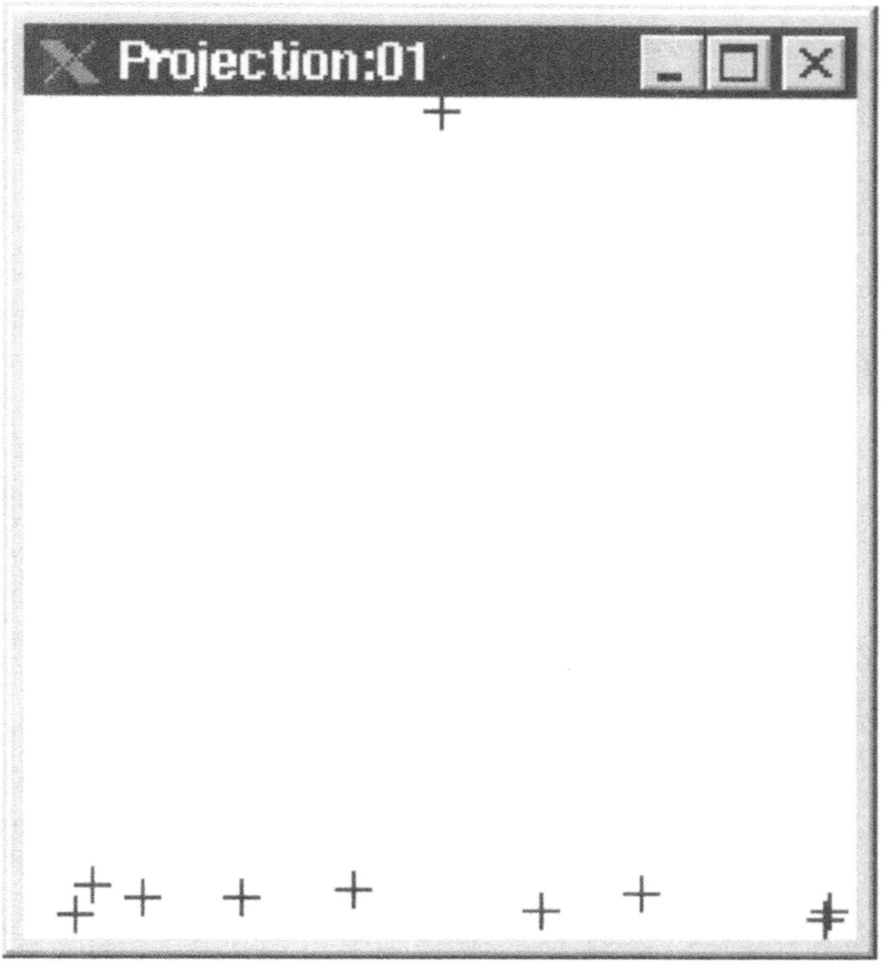

Figure 6.5. Projection 1 (shows how $f(x)$ depends on the first variable).

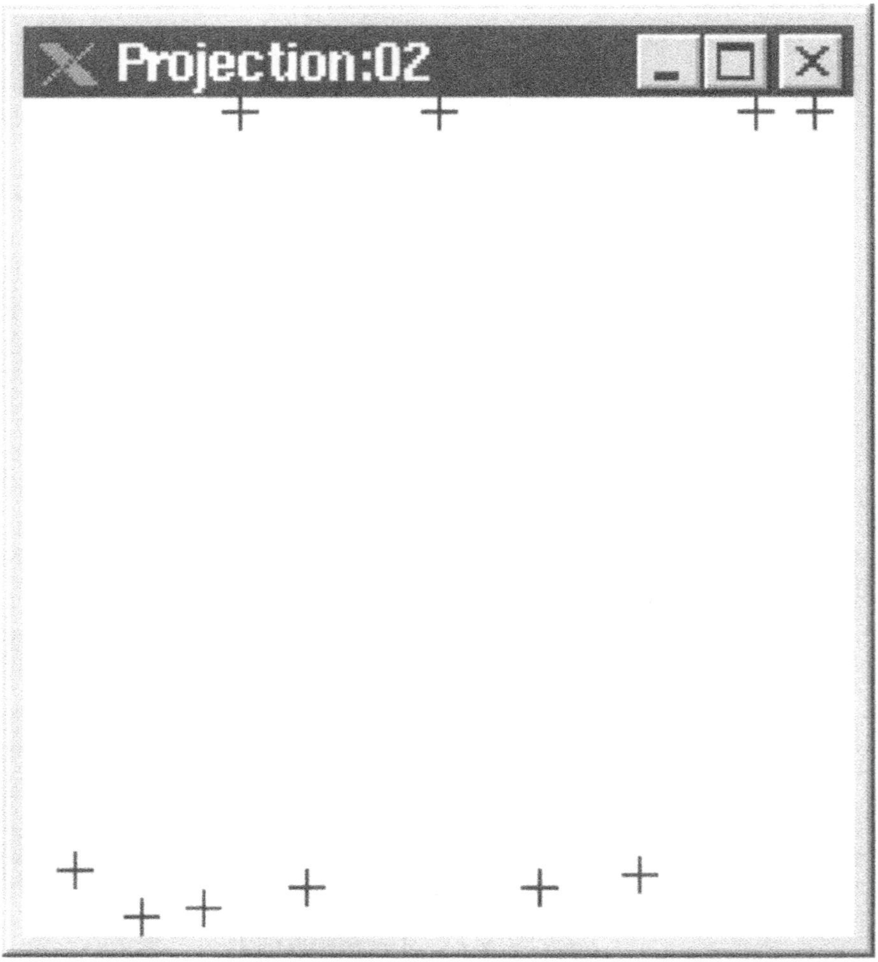

Figure 6.6. Projection 2 (shows how $f(x)$ depends on the second variable).

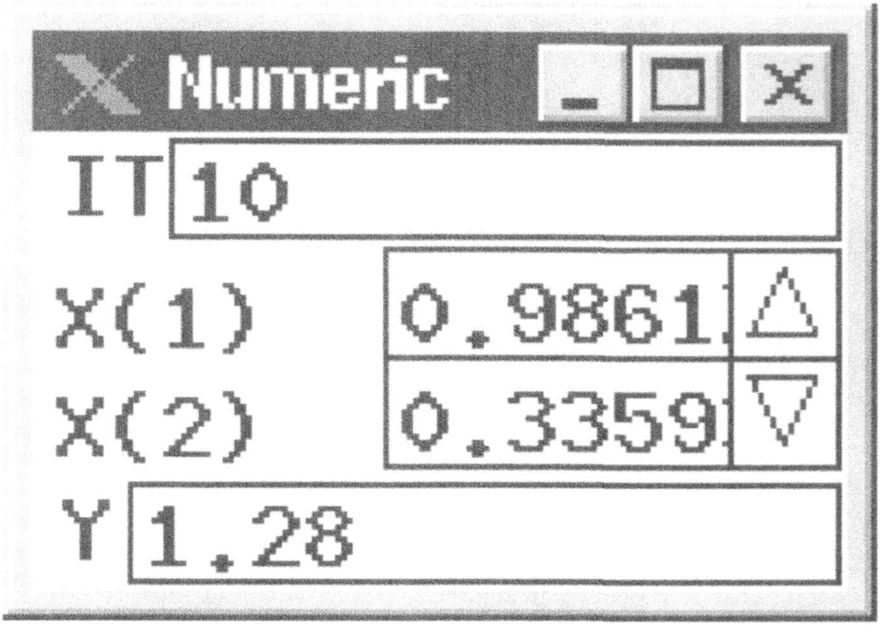

Figure 6.7. Current numeric results.

Chapter 7

JAVA JDK1.0 VERSION (GMJ0)

1. INTRODUCTION

The Global Minimizer GMJ0 (Krapauskas, 1997) implements in Java JDK1.0 the C++ global optimization software GMC. That makes GMJ0 familiar to GMC users. GMJ0 includes all the global methods. Lbayes is the only local method. Implementation of the remaining two local methods, Nlp and Flexi, is difficult. We need to introduce specific 'Constraints' objects[1] to represent constraints, in these methods. The graphical interface and the menu system both are similar to those of GMC.

All 'java' and 'class' files are in the archive on web-sites (see section 4.). The applet 'gm.html' is started by a browser. The command

```
appletviewer (URL address)gm.html
```

starts the applet, if JDK1.0.x[2] is on, and one may reach the corresponding URL address. The applet optimizes the objective function defined by 'Task.java'. In the demonstration applet the 'Task.java' is *sin*.

Users replace an illustrative function in the file 'Task.java' by his own objective function. Then the new 'Task.class' is made by the 'javac'

[1]These objects are implemented in the JDK1.2 version (GMJ2).
[2]The command 'appletviewer' may not work properly, if one uses JDK1.1 or higher. That is a disadvantage of GMJ0.

command. Graphical possibilities of GMJ0 include the convergence, projection and numeric parts, similar to those in GMC.

2. RUNNING GMJ0

2.1 DEFINING OBJECTIVE

The file 'Task.java' represents the objective function. Figure 7.1 shows the illustrative example with objective function *sine*. The file 'Task.java' describes the Task class. The class variable NUMBER OF VARIABLES defines the number of optimization parameters. The arrays 'lowerBounds' and 'upperBounds'[3] define lower and upper bounds of optimization parameters. In the illustrative example, that means variation of parameters from minus two to plus two.

The array 'initialPoints' defines the initial approximation needed by local methods and by some global ones, for example by 'Exkor'. The numbers −1.0 0.0 1.0 define default values of arrays 'lowerBounds', initialPoints', and 'uppperBounds' correspondingly[4]. The 'return' value is the objective function of two variables (see the illustrative example Figure 7.1).

2.2 COMPILING AND RUNNING

One compiles when changes the objective function $f(x)$. This function is defined by in the file 'Task.java'. The compiling command is 'javac Task.java'. To run the program by a local java JDK1.0.x interpreter, one uses the command 'java Gm'. The JDK1.0.x command 'appletviewer gm.html' is for running the applet 'gm.html' from the command line. To run by browser just open the web-site and start the applet 'gm.html' of the GMJ0 system (see section 4.).

[3] Zero element defines the first element, and so on.

[4] Note that the constructor of class Task can be deleted if one is satisfied by the default values.

```
import java.lang.Object;
              // Task
 class Task extends Object
{
// Number of variables
public static final int NUMBER_OF_VARIABLES = 2;
public double lowerBounds[];
public double upperBounds[];
public double initialPoints[];

public Task()
{
// Initialize upper and lower bounds or initial points
 here:
lowerBounds = new double[NUMBER_OF_VARIABLES];
upperBounds = new double[NUMBER_OF_VARIABLES];
initialPoints = new double[NUMBER_OF_VARIABLES];
for ( int i = 0; i < NUMBER_OF_VARIABLES; i++ )
{
lowerBounds[i] = -5.0;
upperBounds[i] = 5.0;
}
initialPoints[0] = 4;//2;
initialPoints[1] = 0;//-4;
}
static public final double f(double x[], int n)
{
  return Math.pow(x[0],2)+x[1];
}
}
```

Figure 7.1. Example of the 'Task.java.'

2.3 GRAPHICAL INTERFACE

Starting the applet 'gm.html', a window with 'Show' and 'Hidden' buttons appears (see Figure 7.2). One starts the main program and opens the menu window (see Figure 7.3) by clicking the 'Show' button. The menu window controls the optimization process and is similar to that of the GMC (C++ version). To hide this window press 'Hide'. To delete, select the menu item 'File, Exit' and answer "yes." Then the 'Restart' button appears, replacing the 'Show' and 'Hide'. One creates

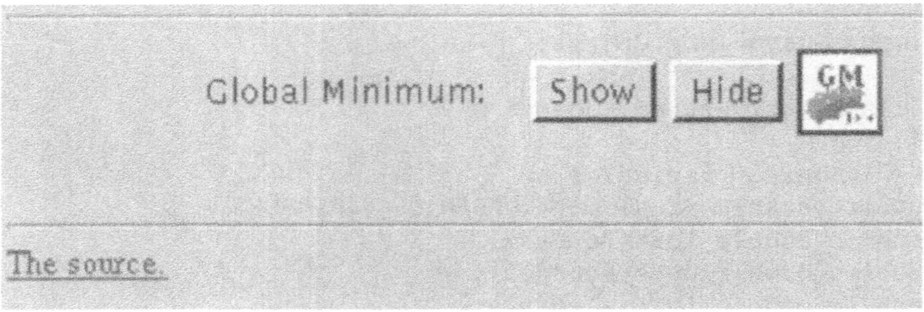

Figure 7.2. GMJ0 opening.

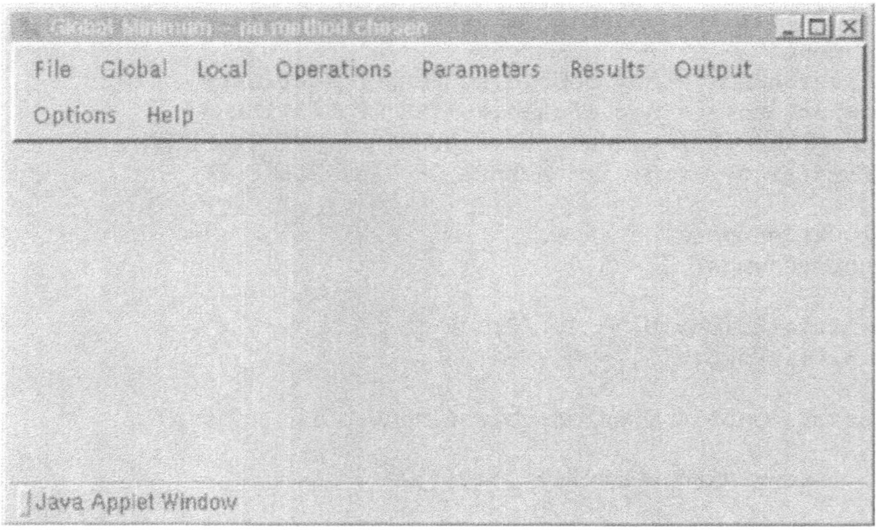

Figure 7.3. GMJ0 opening menu.

a new menu window by clicking 'Restart'. One opens the menu window from the local interpreter by the command 'java Gm.' In this window,

- *File* provides the *Exit* possibility.
- *Global* provides the choice of global optimization methods:
  *Bayes*1, *Unt*, *Migl*, *Exkor*, *Glopt*, *Lpmin*.
- *Local* provides the choice of local methods:
  Nlp, *Flexi*, *Lbayes* (now only *Lbayes* is there).

- *Operation* is for the computing control:
  Run is for starting or restarting, *Stop* is for temporary stopping, and *Terminate* ends the computing.

- *Parameters* is just for *Set initial parameters* that opens the parameter window. The window shape depends on the method chosen. One may change the parameters at any moment but the new settings will be ignored until the next 'run'.

- *Results* is for *Watch Results.* It opens the results window showing the best obtained objective function and the corresponding parameters.

- *Output* is for choosing a visualization format.

  – *Numeric* shows in numbers the current objective and parameters (see Figure 7.4).

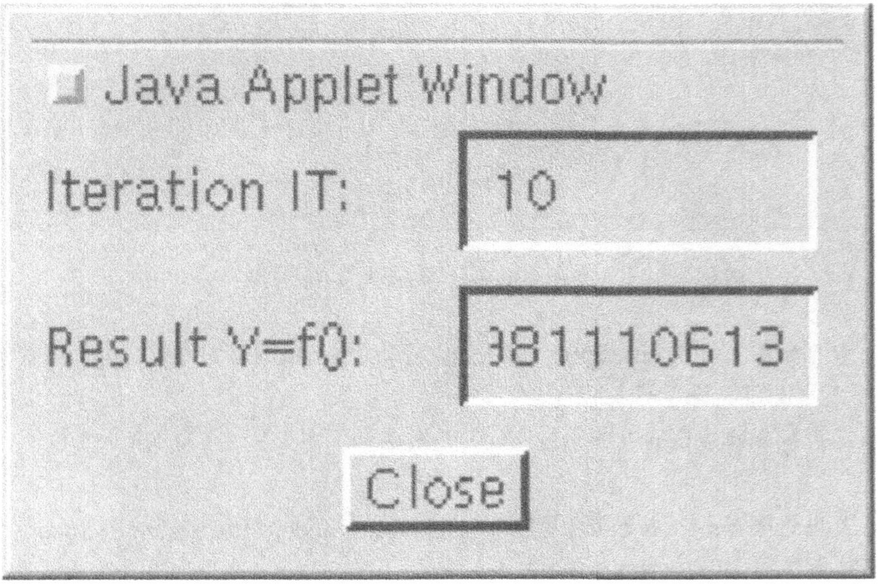

Figure 7.4. Current results.

  – *Convergence* shows how the best obtained objective value depends on the iteration number (see Figure 7.5).

  – *Projection* shows how the objective depends on some fixed parameter (see Figures 7.6 and 7.7).

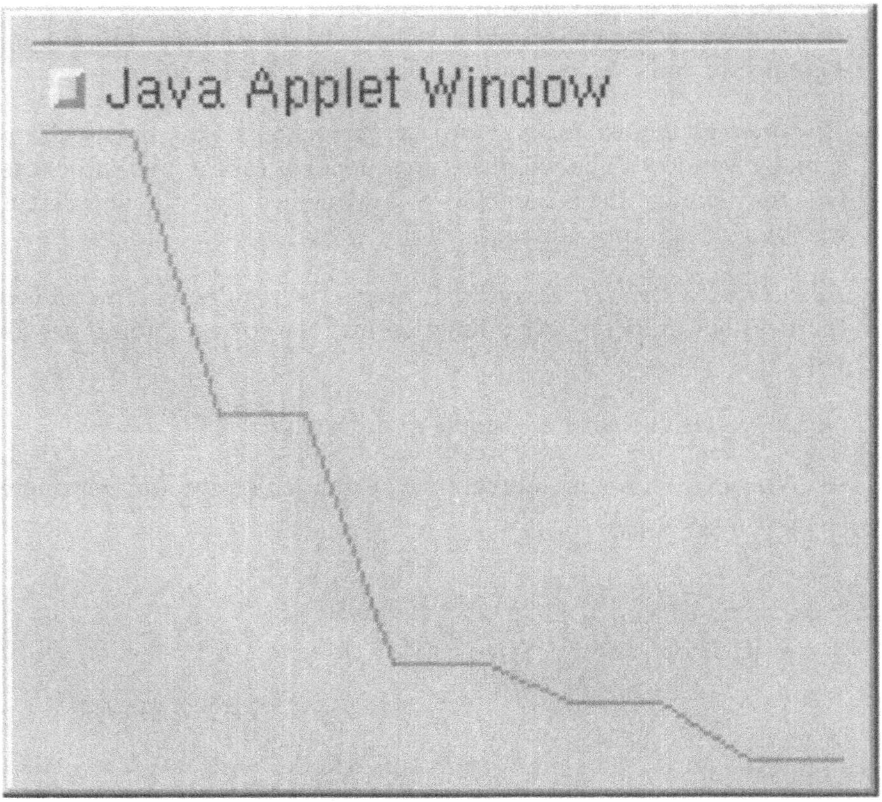

Figure 7.5. Best objective depending on iteration number.

- *Options* is for changing the form of points on the *Projection* window: *Cross points* or *Circle points*.
- *Help* just opens the window *About*.

All the visual tools takes CPU time. Therefore, closing them speeds up the computing.

2.4 IMPLEMENTING NEW METHOD

The class of a new method is created using the 'Method' class. The minimal class of a new method is shown in Figure 7.8. The parameters of the 'Method' constructor *output* is the object controlling a visual output. n defines computing parameters (lower and upper bounds, initial points

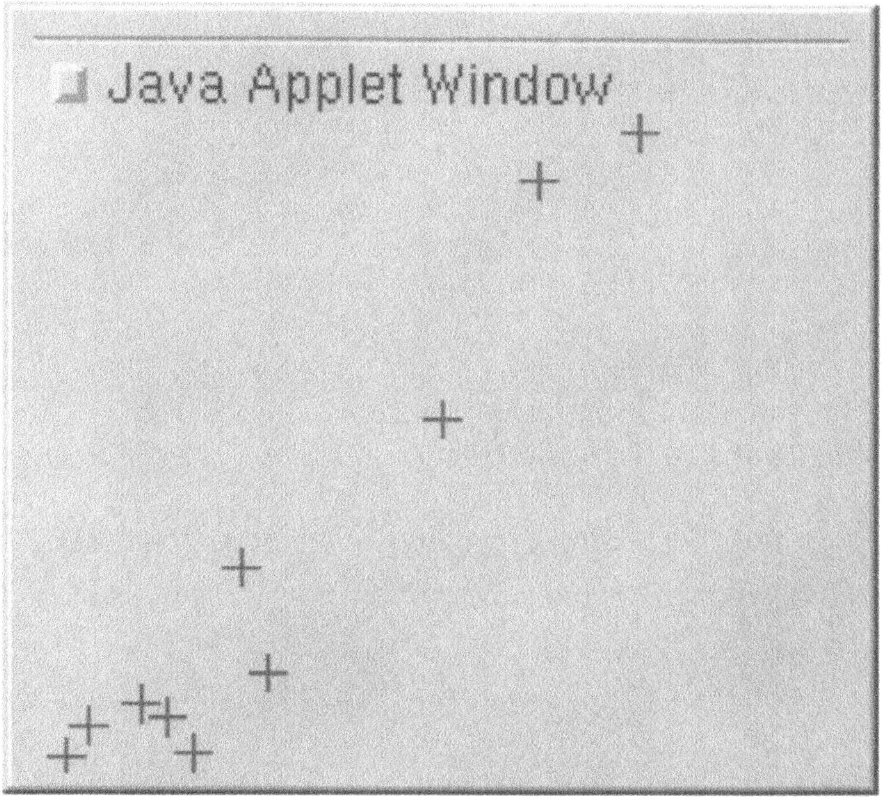

Figure 7.6. Projection 1.

e.t.c.). r defines the results. fm and xn define the arrays of current results.

An optimization algorithm is called by the method 'run()'. In the example, the method is 'newmethodcalculations (...)'. The name and parameters of this method may be fixed according to users needs. In the example, the internal values x, a, and b of the object created by the class 'Method' are passed. Here x, a, and b are some arrays of initial points, lower and upper bounds, respectively. n is the number of variables. it is the number of iterations. fm and xn are arrays of current results.

All these parameters are initialized by the constructor of the 'Method' class. In the 'Method' class, the internal variable r of the class 'Results' is defined. This variable is initiated by the constructor of the 'Method' class. Using the variable r, we inform the program about the

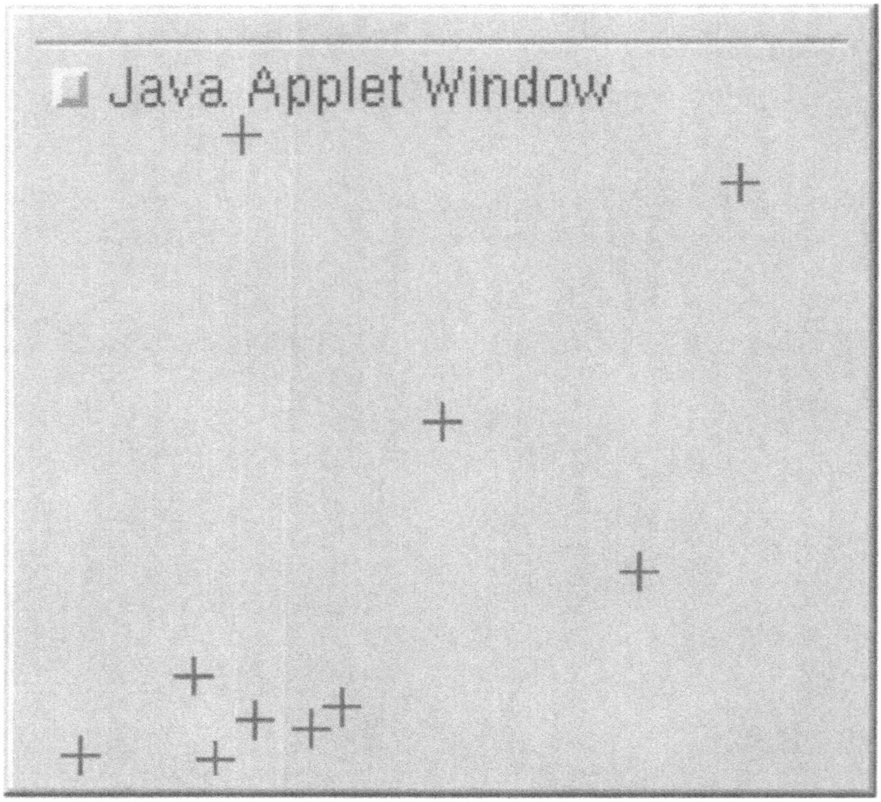

Figure 7.7. Projection 2.

change of iteration (r.iteration=new.iteration), about new minimal value (r.y=new.min), about the change of the current best point (copy.array(r.x=new.x.n)). Here the integer 'new.iteration' shows the iteration number. The double array "new.min' defines the coordinates of the new best point. All these commands are in the method 'newmethod-calculations'. Finishing the method 'run()', the method 'Finish()' of the class 'Method' is called. Then the program returns to the initial state.

One may add a new method and extend the graphical possibilities of Java global optimization software in a simpler way by using other Java versions. These versions called as GMJ1 and GMJ2 are described in the next section. Note, that GMJ1 needs JDK1.1 support and GMJ2 needs JDK1.2 support. Thus, GMJ1 does not work on some older browsers such as Netscape-3, for example. One installs the Java Developers Kit JDK1.2 to apply GMJ2.

```
import Method;

class NewMethod extends Method
{
 public NewMethod( Output output, Parameters p, Results r,
 double fm[], double xn[][] )
 {
  super( output, p, r, fm, xn );
 }

 public void run()
 {
  newmethodcalculations(x, a, b, n, it, fm, xn);
  finish();
 }

 private void newmethodcalculations(double x[], double a[],
 double b[], int n, int it, double fm[], double xn[][])
 {
 //code
 }
}
```

Figure 7.8. Example of the minimal class of the new method.

Chapter 8

JAVA JDK1.1 AND JDK1.2 VERSIONS, GMJ1 AND GMJ2

1. INTRODUCTION

The Java JDK1.1 and JDK1.2 versions (in short, GMJ1 and GMJ2) of global optimization software are class frameworks (Grybauskas, 1998). Both GMJ versions implement:
global optimization algorithms ($METHODS$),
functions to be optimized ($TASKS$),
objects to be represented visually ($ANALYSIS$).

GMJ can be run as an applet or as a stand-alone application. One can run applet over the Web. This cannot be done by an application. Security restrictions is the disadvantage of applets. We need security permits to read/write files and to connect third parties (see Section 4.6).

These restrictions are lifted using stand-alone applications. Using stand-alone applications, the most recent Java features supported by the Java Development Kit (JDK) can be exploited. The reason is that web browsers are late adapting new Java Developer Kits (JDK), as usual. If JDK is installed, then applets can be loaded directly by appletviewers[1].

[1] For example, by the command
appletviewer http : //optimum.mii.lt/ jonas/gmj2j/gmj.html.

The JDK1.2 system provides new facilities as compared with JDK1.1. For example, Just-in-Time compilers (jitc) increase the speed several times[2]. The 'policy' files of JDK1.2 exempts trusted users from some security restrictions.

2. STARTING GMJ

In this and flowing sections, the user guide of GMJ is described. First we consider the parts that are common for both JDK1.1 and JDK1.2. Then the features of JDK1.2 are mentioned.

2.1 CONFIGURING APPLET

The sample file *gmj.html* shows how to set configuration parameters of the *gmj* applet. The < *applet* > tag of GMJ1 looks like this:

```
<html>
<head>
   <META HTTP-EQUIV="Content-Language"  CONTENT="LT">
    <META HTTP-EQUIV="Content-Type" CONTENT= "text/html;
charset=windows-1257; charset=ISO-8859-13">
    <META NAME="Author" CONTENT="Modestas Grybauskas">
 <title>GMJ Applet,    Set of Methods, Tasks, and Analysis
Objects</title>
 </head>
 <body bgcolor="#E0E0E0">
 <h1>GMJ</h1>
<p>
<applet
 code="GMJext.class"
  codebase= "Lib" height=1200 width=800
  archive= "gmjg_allnew.jar"
 >
 <patram name="TASKS"
value="lt.ktu.gmj.tasks.Sin|
 Tvarka|lt.ktu.gmj.tasks.Portfolio|
lt.ktu.gmj.tasks.WalrasModel|lt.ktu.gmj.tasks.Knapsack">
<param name="METHODS" value="lt.ktu.gmj.methods.Mig1|
lt.ktu.gmj.methods.Bayes|lt.ktu.gmj.methods.Globt|
```

[2]These compilers are available in JDK1.1, too, starting from JDK1.1.7.

```
lt.ktu.gmj.methods.Unt|lt.ktu.gmj.methods.LBayes|
lt.ktu.gmj.methods.Exkor|lt.ktu.gmj.methods.Flexi|
lt.ktu.gmj.methods.Lpmin">
<param name="ANALYSIS"
 value="lt.ktu.gmj.analysis.Convergence|
lt.ktu.gmj.analysis.Spectrum|
lt.ktu.gmj.analysis.Projection|
TvarkaAnalyser|lt.ktu.gmj.analysis.WalrasProfit">
  <p>You need Java compatible browser
 to see this applet </p>
 </applet>
  </p>
 </body>
 </html>
```

In this example, *code* specifies the applet class and should not be changed.
codebase specifies the relative URL of the applet class archive file and user class files. The URL is relative to the location of the HTML file where the $<$ *applet* $>$ tag resides or is an absolute path.
archive lists the class archive files.
width and *height* specifies the applet size, as it appears on the HTML page.
TASKS lists tasks which are available when the applet loads. Complete class names (package and class name) are separated by | symbol.
METHODS lists methods which are supported when the applet loads.
ANALYSIS lists visual analysis classes which are supported when the applet loads.

Here is the fragment of the GMJ2 $<$ *applet* $>$ tag, which differs from GMJ1:

```
<applet
 code="lt.ktu.gmj.ui.GMJ.class"
  codebase= "Lib"
  archive= "gmj20.jar"
 >
 <patram name="TASKS"
value="lt.ktu.gmj.tasks.Sin">
<param name="METHODS" value="lt.ktu.gmj.methods.Bayes">
<param name="ANALYSIS"
 value="lt.ktu.gmj.analysis.Convergence|
lt.ktu.gmj.analysis.Spectrum|
```

```
lt.ktu.gmj.analysis.Projection">
 </applet>
```

The difference is that less methods, tasks and analysis objects are included in the GMJ2 example. Different archive file *gmj20.jar* is used.

2.2 CONFIGURING STAND-ALONE APPLICATION

To run GMJ as a stand-alone application, the Java Runtime Environment (JRE) is required. First, *CLASSPATH* should be configured, so that the GMJ classes can be found by the JRE. An application is started by loading the *lt.ktu.gmj.ui.GMJ* class. Using Linux the command is:

```
java -classpath .
:/usr/local/jdk1.2/lib/classes.zip:
/home/jonas/public_html/gmj2j/Lib/gmj20.jar
lt.ktu.gmj.ui.GMJ  lt.ktu.gmj.tasks.Sin
 lt.ktu.gmj.methods.Bayes
lt.ktu.gmj.analysis.Projection
```

The application accepts three optional parameters: tasks, methods, and analysis objects. In the example, only one task *Sin*, one method *Bayes* and one analysis object *Projection* is included, for simplicity.

3. RUNNING GMJ

3.1 DISPLAY AND CONTROL

GMJ displays a tab control that has three choices: *method*, *task*, and *operation*. The appropriate pages can be selected by clicking on the page tabs.

Figures 8.1,8.2, and 8.3 show the method , the task and the operation pages for the *Exkor* method and the *Sin* task using GMJ1.
Figure 8.4 shows how the function depends on the first variable x
Figure 8.5 shows how the function depends on the second variable y.

Figure 8.1. Method page with parameter fields.

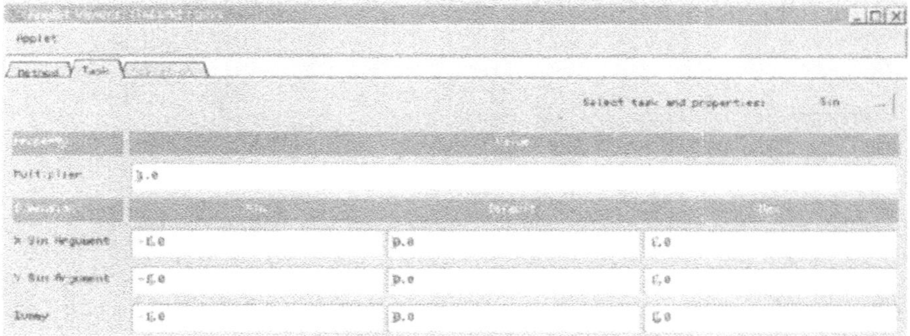

Figure 8.2. Task page with input fields.

Figure 8.3. Output page with results fields.

Figures 8.4 and 8.5 show that using the *Exkor* method one gets clear projections defining how the objective function values depend on different variables. That provides a sort of "visualization" needed for the preliminary investigations. Using other methods, *Projection* windows show fuzzy "projections." The reason is that variables change together.

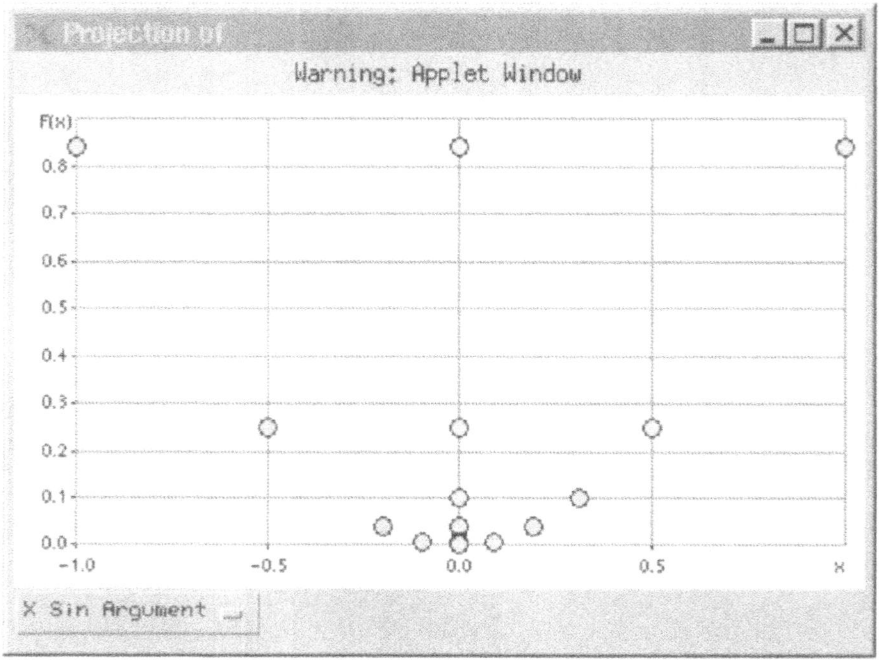

Figure 8.4. x-projection.

Figures 2.3 (bottom), and 2.4 show projections obtained while optimizing the randomization parameters for the knapsack problem by the method *Bayes* .
Figure 2.3 (top) shows how the best current value of the knapsack depends on iteration numbers.
The *Spectrum* window[3] shows how depends the frequency of the objective function $f(x)$ values on iteration numbers. Different frequencies are represented by different colors.

3.2 TASK SELECTION

The following tasks are installed in GMJ1.

Sin : the Sine Function.

[3]This window can be open by selecting the *Spectrum* mode on the *Convergence* menu.

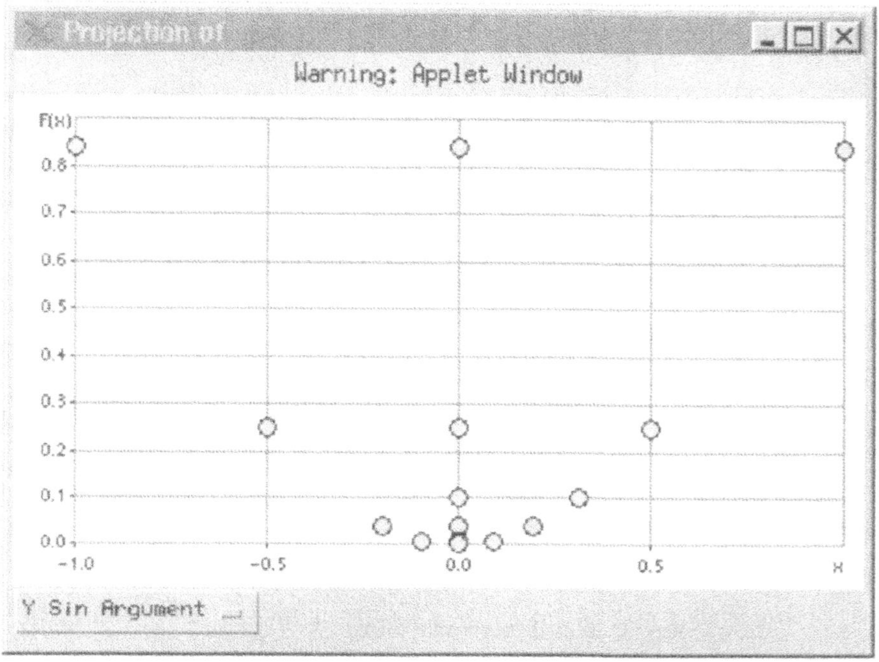

Figure 8.5. y-projection.

Tvarka : the School Scheduling.

Portfolio : the Optimal Investment, the "Portfolio" problem.

WalrasModel : the Walras Equilibrium.

Knapsack : the Knapsack problem.

The detail description of these examples is the subject of the next part of this book. By clicking the *tasks* label, one opens a list of available tasks. The task is selected by clicking the label of this task.

Figure 8.2 shows the selected task *Sin*. After selection, the tasks implementation class is loaded and the configuration window is opened. The configuration window depends on the actual task selected.

Figure 8.2 shows the window for *Sin*. There are fields that define properties of tasks and the minimization domain. One can keep the default values or change them. If the entered value is invalid or out of bounds, the field turns red and the value is not stored.

3.3 OPERATION CONTROL

The minimization process is controlled by the operation page (see bottom of Figure 8.3). There are three control buttons: **run, pause, stop**. The **run** button starts the optimization. The **pause** button interrupts the calculations. To abandon them completely, press the **stop** button.

3.4 ANALYSIS OF RESULTS

The numerical optimization results are shown in the output table on Figure 8.3 displaying the following parameters:

Iteration shows current iteration numbers,

$F(x)$, shows current values of the objective function $f(x)$,

X, Y , show coordinates of current points.

To select the form of graphical results, one clicks the 'Convergence' field that is at the right side on the top. Three analysis objects are installed.

Convergence shows how the best obtained objective function value depends on iteration number (see Figure 2.3),

Projection shows how the objective function depends on the first variable (see Figure 8.4), the variable number can be changed by clicking at the small button in the lower-left part of the window,

Spectrum shows how depends the frequency[4] of objective function values on the iteration number.

3.5 METHOD SELECTION

In GMJ1 five global methods are installed:

Migl : Monte Carlo.

[4]Defined by different colors.

Bayes : Bayesian by Mockus.

Unt : Extrapolation by Zilinskas.

Globt : Clustering by Torn.

Exkor : Bayesian coordinate search by Zilinskas.

and two local methods:

LBayes : Stochastic Approximation by Mockus.

Flexi : Simplex Nonlinear by Nelder and Mead[5].

For the detail descriptions see (Mockus et al., 1997).

One opens a list of available methods by clicking the *methods* label of the GMJ tab control. The method is selected by clicking the label of this method. Figure 8.1 shows the selected method *Exkor*. After the selection the method's implementation class is loaded and the configuration window is opened.

The configuration window depends on the actual method selected. As usual, there is the *Iterations* field where the number of iterations is defined, by keeping the default values or by changing them. The field *Initialpoints* (see Figure 8.1) shows the number of observations for a preliminary survey. The preliminary survey is done by some methods including *Bayes*. If the entered value is invalid or out of bounds, the field turns red and the value is not stored.

4. UPDATING GMJ

Compiling is needed to add new methods, new tasks, new analysis objects. One compiles, too, to change the data which is not on the graphical interface. For example, to add a new task 'NewTask.java', a new method 'NewMethod.java', and a new analysis object 'NewAnalysis.java' one does these operations:

[5]This is a reduced Flexi version. Only rectangular constraints are implemented.

- compile the 'NewTask.java' , 'NewMethod.java', and 'NewAnalysis.java' making the corresponding *class* files

```
javac  -verbose
 -g -classpath .:
/home/jonas/public_html/gmjgallj
/Lib/gmj_allnew.jar:
/usr/local/jdk1.1.3/lib/classes.zip
/home/jonas/public_html/gmjgallj
/lt/ktu/gmj/tasks/NewTask.java
/home/jonas/public_html/gmjgallj
/lt/ktu/gmj/methods/NewMethod.java
/home/jonas/public_html/gmjgallj
/lt/ktu/gmj/analysis/NewAnalysis.java
```

- create the additional archive file '*gmjg_add.jar*' by including the new class files

```
 jar -cvf   /home/jonas/public_html/gmjgallj
 /Lib/gmjg_add.jar
*class lt/*
```

- update the applet 'gmjall.html' by updating the *TASKS*, *METHODS*, and *ANALYSIS* parameters and by adding the new archive file

```
<html>
<head>
   <META HTTP-EQUIV="Content-Language"
CONTENT="LT">
   <META HTTP-EQUIV="Content-Type" CONTENT=
"text/html; charset=windows-1257; charset=ISO-8859-13">
   <META NAME="Author" CONTENT="Modestas Grybauskas">
<title>GMJ Applet,
Set of Methods, Tasks, and Analysis Objects</title>
</head>
<body bgcolor="#E0E0E0">
<h1>GMJ</h1>
<p>
<applet
 code="GMJext.class"
codebase= "Lib" height=1200 width=800
```

```
archive= "gmjg_allnew.jar  gmjg_add.jar"
 >
<patram name="TASKS"
value="lt.ktu.gmj.tasks.Sin|
Tvarka|lt.ktu.gmj.tasks.Portfolio|
lt.ktu.gmj.tasks.WalrasModel|lt.ktu.gmj.tasks.Knapsack|
lt.ktu.gmj.tasks.NewTask">

<param name="METHODS" value="lt.ktu.gmj.methods.Migl|
lt.ktu.gmj.methods.Bayes|lt.ktu.gmj.methods.Globt|
lt.ktu.gmj.methods.Unt|lt.ktu.gmj.methods.LBayes|
lt.ktu.gmj.methods.Exkor|lt.ktu.gmj.methods.Lpmin|
lt.ktu.gmj.methods.Flexi||lt.ktu.gmj.methods.NewMethod">

<param name="ANALYSIS"
 value="lt.ktu.gmj.analysis.Convergence|
lt.ktu.gmj.analysis.Spectrum|
lt.ktu.gmj.analysis.Projection|
TvarkaAnalyser|lt.ktu.gmj.analysis.WalrasProfit|
lt.ktu.gmj.analysis.NewAnalysis">
 <p>You need Java compatible browser
 to see this applet </p>
</applet>

</p>
</body>
</html>
```

4.1 NEW TASKS

The task class implements the Task interface:

```
public interface Task extends Customizable
{
Domain domain ();
double f (Point pt);
};
```

The interface *Task* is derived from *Customizable.* This means that it is configurable by users. Details are explained in Section 4.3.
A user defined method *doublef* calculates values of the objective function at the *Point pt*

The class *Point pt* defines the array of optimization variables belonging to the *Domain*.
The object *domain* returns the class *Domain*, which defines a set of feasible decisions.

In GMJ1 this is a rectangular region defined by the lower and upper bounds of optimization variables[6].
The *Domain* is an abstract class:

```
public abstract class Domain implements Customizable
{ public double min[]; public double max[];
 public Point defaultPoint;
public abstract String[] dimensions (); ..... };
```

The abstract function *dimensions* should be overridden to return the descriptions of the domain dimensions. These descriptions are in analysis and configuration windows.
Variables *min*, *max* and *defaultPoint* are created by the class *Domain* and initialized by the derived class, as required. The *Point* class implements a point in a n-dimensional set *Domain*.

```
public class Point { public double x[]; public Point
 (Domain domain) ... };
```

The class *Point* constructs from a *Dimension* object.

One does the following six steps to implement a new task:

- derive a class from the class *Domain*,
- provide the constructor to initialize the default dimension range and the default point,
- override the method *dimensions* and return descriptions of the dimensions of the domain,
- create the new class, which implements the interface *Task*, or extend an abstract class AbstractTask to make the code compatible with future GMJ releases,

[6]Note, that in GMJ2 the feasible set is given by a user defined *constraint* function. Therefore, there are no restrictions on the shape of feasible set.

- override the method *domain*,
- override the method *f*.

Example of Task Sine.

```
class SinDomain extends Domain
{
static final String [] dimensions=
{"X Sin Argument"};
public String [] dimensions ()
{return dimensions
}
SinDomain ()
{
min [0]=-2;
max [0]= 2;
defaultPoint.x[0]=0;
}
};
public class Sin implements Task
{
private SinDomain domain=new SinDomain ();
public Domain domain ()
{return domain;}
public double f (Point pt)
{
return Math.sin (pt.x[0]);
}
};
```

4.2 NEW METHODS

GMJ can be extended providing additional optimization methods, which may be more efficient or more flexible. The method class should implement the *Method* interface:

```
public interface Method extends Customizable
{
int iterations ();
Result run (ResultLogger  l, Task t);
};
```

The interface *Method* is derived from the interface *Customizable*, which requires it to be configurable by the user. *iterations* returns the total number of iterations.

Most of the optimization methods have the configurable *iterations* property. Therefore, the *MethodBase* abstract class is implemented on top of the *Method* interface. It overrides the *iterations* and the *customize* methods. The method *run* is overridden by inheriting from *MethodBase*. The method *run* runs the optimization method on the selected *Task*, writes the results to the specified *ResultLogger*, and returns the best result.

The *ResultLogger* interface gathers the generated results and stores them. The *ResultLoger* interface defines the following members:

```
public interface ResultLogger  { void log (Result r);
ResultlogAt (int it); };
```

The method *log* is to log the result of each iteration of the method. Usually, the class implementing the *ResultLoger* updates the user interface objects reflecting the method's progress when *log* is called. Some methods need a history of optimization. The method *logAt* provides that by returning the *Result* logged previously in the iteration *it* using the method *log*. The class *Result* contains the iteration state:

```
public final class Result  {
public int iteration; public Point
point; public double value; ..... };
```

4.3 CONFIGURING PROPERTY MANAGER

Many objects are configured by users. For example, a user may want to change the number of iterations or the ranges of the optimization domain. *PropertyManager* shows the properties, which may be changed. The *PropertyManager* is a property list that is described by this interface:

```
public interface PropertyManager
{
void removeAll ();
void add (Property property);
```

```
void validate ();
};
```

The method *add* is to add new *Property* objects to the list. The property consists of a property name or a label and a value editor. The value editor is for changing the value of the property.

The important component of the *Property* object is the *PropertyProvider*. This is an object, which knows how to retrieve the value of the property and how to store it after the changes made by the editor. The custom value retrieval and storage methods are implemented because in Java one cannot to pass primitive objects by reference.

A set of property providers is in the GMJ package, including:

- the class *DoubleArrayProvider* accesses the double array elements
- the class *FieldProvider* uses the JDK1.1 Reflection classes providing access to any field of a class

Example. The following sample code shows how to add a property called the "Initial Speed" stored as the second element of the array *customData*. The initial speed can change in the range from 10 to 50 km/h. by this interface:

```
manager.add
{
new DoubleProperty
{"Initial Speed (km/h)",
new DoubleArrayProvider (customData, 1), 10, 50 }
};
```

Here is an example how to add a property named "Offset" stored in a public variable *offset* of the calling class

```
manager.add
{
new DoubleProperty
{"Offset",
new FieldProvider (this, "offset"), 0, 1 }
```

```
};
```

Adding a property to the property manager, an editor of the appropriate type is created. If the user changes the value, the value is automatically checked, validated and stored in the second element of the array. *customData*.

WRITING CUSTOM PROPERTY PROVIDER

When a property value is stored as a field, then a custom access class should be provided to reach that value. This class implements the *PropertyProvider* interface requiring many functions to be overridden. Most of them are implemented in the *SimplePropertyProvider* class, which can be used as a base.

Example. This sample code shows how to provide an access to the field
multiplier of the *Sin* class, which is *double*:

```
class SinMultiplierProvider extends SimplePropertyProvider
{
public Sin sin;
public SinMultiplierProvider (Sin _sin)
{sin = _sin;}
public Object get ()
{return new Double (sin.multiplier);}
public void set (Object value)
throws InvalidPropertyException
{sin.multiplier=((Double) value).doubleValue ();}
};
```

Note, that in Java 1.2 reflection is always available. Therefore, applying GMJ2 one applies *FieldProvider*, instead of writing custom providers.

4.4 NEW ANALYSIS OBJECTS

GMJ provides three analysis objects: *Convergence*, *Projection* and *Spectrum* (see section 3.4). The new tasks may need the new analysis objects. The new analyzers may be derived from any base class. They can be implemented as some windows, or they may log data to some external files. The only requirement is to implement the following *ResultAnalyser* interface:

```
public interface ResultAnalyser
{
void prepare (ResultRepository r);
void dispose ();
};
```

To construct analysis objects the method *prepare* is called. It asks the analysis objects to initialize. A *ResultRepository* object is passed as a parameter. It contains information to initialize the analyzer.

The method *dispose* is called when the analyzer is being deleted. It flush the open buffers or close associated windows. The data observer-notifier pattern is to deliver the results of real time calculations to the analysis objects. The data notifier has a list of data observers, which are notified when new data is available. To receive these notifications, the object implements the *ResultObserver* interface and registers on the appropriate *ResultRepository* class. The registration time is in the *prepare* method: before leaving.

```
public interface ResultObserver
{
void newResult (ResultRepository l);
void newProgressResult (ResultRepository l);
void finalResult (ResultRepository l);
void newLog (ResultRepository l);
void oldLog (ResultRepository l);
};
```

The method *logStarted* signals that new calculations are going to be started.
The method *newResult* signals, that a new result has just been added to the *ResultLoger*.
The method *newProgressResult* signals that a method has made progress and did find the better point.
The method *finalResult* signals that the optimization is over.
The method *logDisposed* signals that the *ResultLog* is depreciated and

cannot be used anymore. In such a case the observers should delete all the references to the *ResultLogger* objects.

Example. Here is a simple analyzer to print the progress to console windows

```
public class Demo
 ResultObserver  implements ResultAnalyser,
ResultObserver {
protected ResultRepository results=null};

public void prepare {ResultRepository n}
{
results=n;
results.addObserver {this};
}

public synchronized void logStarted (ResultRepository n)
{
System.out.println {"START"};
}
void newResult (ResultRepository 1) {}
void finalResult (ResultRepository 1) {}

void newProgressResult (ResultRepository 1)
{
System.ot.println {"Progress was made at iteration "
+ 1.lastResult ().iteration);
}

public void logDisposed (ResultRepository r)
{
System.out.println ("STOP");
dispose ();
}
};
```

4.5 PROGRAMMING TIPS

ABORTING CALCULATIONS

If optimization lasts too long, a user wants to abort the calculations, as usual. When the *STOP* button is pressed or the applet is about to exit, then an *interrupted* flag is set for the *calculation* thread. Therefore, *task* and *method* writers are advised to test for

```
if (Thread.interrupted ())
 throw new ResultLogger.AbortedEror ();
```

IMPLEMENTING INTERFACES

Creating a custom method or task, the class implements a corresponding interface, as usual. The better alternative is to extend an abstract class, if available. For example, "extend *AbstractMethod*" is better as compared with "implement *Method*." This makes a code more compatible with future GMJ releases.

URL ADDRESS

The original web-sites are attached to the servers 'optimum.mii.lt' or 'mockus.org'. Therefore, in some examples, the URL addresses should be updated to extract data from the local server .

On the 'Task' menu of GMJ1 global optimization software the examples are: "Portfolio," "Tvarka," and "Knapsack." In the task of school scheduling 'Tvarka.java', the URL address of the applet host is defined automatically and is shown on the graphical interface. In the task "Knapsack", some URL address is shown on the screen. It should be updated by users. In the "Portfolio" task, the URL is defined in the file 'Portfolio.java'. Therefore, recompiling is needed to extract the data.

4.6 SECURITY RESTRICTIONS

STAND-ALONE APPLICATION

Loading applets from their hosts, called as "applet hosts," one meets security restrictions. Applying GMJ one often needs to read/write a file on a client site or to connect to a strange host. A host that is different from the applet host is called as the "strange host."

If GMJ is loaded as a stand-alone application, one can accomplish these tasks without security restrictions. Loading GMJ as an applet one is running in a "sand-box." That means: no reading/writing files on client sites, no connections to strange hosts.

SIGNED APPLETS

A way to escape the sand-box is to sign the applet. The signature is a guarantee to others that the applet is friendly. The signature does not prevent damage, but it serves as a stamp of good intention. It identifies the responsible party, if damage does occur.

A way to guarantee that a signature is authentic is to notarize it by a Certificate Authority (CA). The Certificate Authority authenticates the identity of the applicant and then issues a signing certificate.

Anything signed with a signing certificate will bear the name of CA besides the signer name. The signature can be traced, by way of CA, to the signer.

"VeriSign,Inc." is the recognized CA. Signing certificates can be obtained at:
http://digitalid.verisign.com/software_publishers.html
For an individual, it costs $40 per year for both Netscape and Internet Explorer. The prices are expected to drop. Here is a fragment of a signed applet:

```
<applet code="WriteFile.class"
ARCHIVE="signedWriteFile.jar"
<param name="CABBASE"
value="signedWriteFile.cab">
</applet>
```

The ARCHIVE parameter is parsed by Netscape. The CABBASE parameter is parsed by Internet Explorer. Different browsers use different signatures.

For example, the "Netscape" requires to change some parts of Java software, besides signing *jar* files.

Signing in "Sun Appletviewer" is simpler. The description is in the JDK documentation. Here a user needs the certificate and a "policy" file, where all his rights are listed.

To sign "Internet Explorer" one downloads the Microsoft Java Developer Kit. The additional information is on web-sites:
http://www.fastlane.net.~tlandry/javafaq.txt
http://ourworld.compuserve.com/homepages/
jozart/article/index.html

Note, that one should not change the existing 'gmj20.jar' file in GMJ2. Otherwise, security restrictions might be violated. Therefore, extending GMJ a user creates his own *jar* files[7].

5. FEATURES OF GMJ2

5.1 CUSTOM RESULT ANALYZERS

Visualization is a convenient way to represent the input of tasks and the results. As usual, the different tasks need different visualization. For example, considering the school scheduling (see chapter 3.), one wants to see the best obtained schedule. Considering the existence of the equilibrium in the Walras model (see chapter 10), it is important to see how profit functions depend on prices, resources, and other parameters.

In such cases the output is implemented as an analysis object (see section 4.4). However, we need many analysis objects to satisfy all the tasks. To show only task-specific analysis objects, in GMJ2 the *Task* class implements the *yzers0* interface

```
public interface yzers 0
```

[7]In the "applet.html" file one can list several *jar* files.

```
 extends Task
{
Class () analyzers () throws dException 0
;
};
```

The *analyzers* function returns a class list of custom analyzers of the task.

Example. This is a sample task implementation. It states, that there exists a custom analyzer, called "Diagram".

```
class Demo extends AbstractTask implements ers 0

{
.....
public Class () analyzers () throws ClassNotFoundException
{
Class digramClass.forName ("lt.ktu.gmj.demo.Diagram");
return new Class () (diagramClass );
}
...
};
```

5.2 DOMAIN CONSTRAINT FUNCTION

Using GMJ1, the methods consider rectangular feasible regions. These regions are bounded by lower and upper limits defined on the screen. Therefore, in GMJ1 "non-rectangular" constraints[8] are included using penalty functions. This is not always convenient.

In GMJ2 the class *Domain* is extended to represent any non-rectangular domain. This domain includes the function.

```
public double constraintAt {Point pt}
```

[8]By "non-rectangular" we mean constraints defined as a system of linear and nonlinear equations and inequalities.

This function defines a constraint value at the point *pt*. It returns number 1 by default (see Figure 8.6). Writing new methods in GMJ2, one checks the constraint function before calling the objective function f of the *Task*. If a method does not support constraints, one multiplies the constraint value[9] by f and minimizes the result.

Implementing a task in GMJ2, the custom domain class can be derived from the *DomainWithConstraint* class. This enables a user to specify the constraint function at run time. *DomainWithConstraint* adds an entry field in the task configuration panel. In this panel, any algebraic expression can be entered and compiled at run time (see Figure 8.6). Here the constraint is entered by algebraic expression $-X1^2 - X2^2 + 1$ meaning that $-x_1^2 - x_2^2 + 1 \geq 0$. Using these tools, GMJ2 users can specify any constraint function.

5.3 RUNNING GMJ2

The *constraint* function is included into the *Tasks* page of of GMJ2 (see Figure 8.6). In this Figure the *Sinn* task is active. There are several different tasks: *Tvarka*, *WalrasModel*, *DuelStarwars*, *FlowShop* and four types of sine and cosine. *Sinn* is sine with a non-linear constraint. *Sinl* is sine with a linear constraint. *Sinc* is sine without constraints and *Cosc* is cosine without constraints. The task *Tvarka* is for the school scheduling (Nemajunas and Remeikis, 1999). *WalrasModel* is for the Walras problem and *DuelStarwars* is for the Duel problem.

The *Methods* menu of GMJ2 is in Figure 8.7. There are five global methods: *Migl*, *Bayes*, *Exkor*, *Globt*, and *Unt* and three local methods *Shitk*, *Flexi*, and *Lbayes*. The method *Shitk* is the Java version (Margelevicius, 1999) of the non-linear programming method by Schittkowski (Schittkowski, 1985). Names of remaining methods are the same as in GMJ1.

The *Analysis* menu of GMJ2 is in Figure 8.8. Three analysis objects: *Convergence*, *Projection*, and *Spectrum* are common for all tasks. Four analysis objects: *Tvarka*, *WalrasProfit* and *DemoAnalyzer* are task-specific. *Tvarka* shows the initial and the optimized school schedules. *WalrasProfit* shows how the server profit depends on prices and other

[9]The default value is 1.

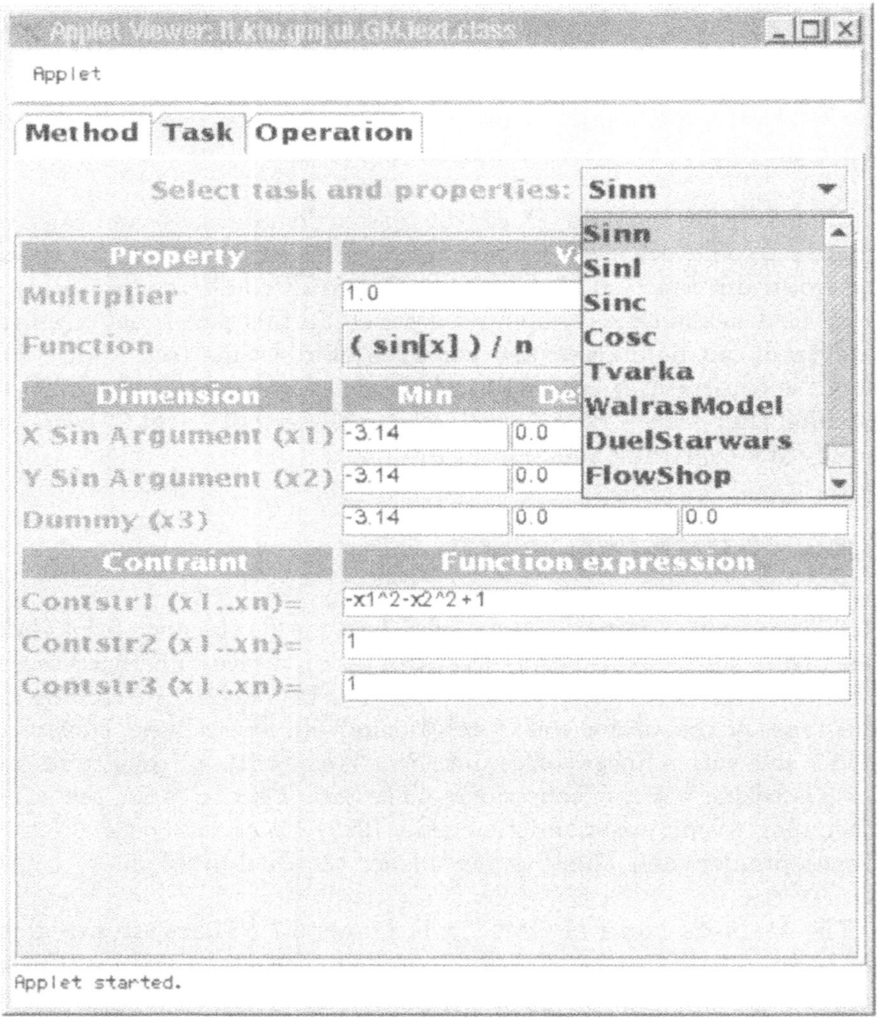

Figure 8.6. Task page for the *Sin* task with one nonlinear constraint.

parameters. The task-specific analysis objects should be seen, if corresponding tasks are on. Otherwise, they should be closed. The analysis object *DemoAnalyzer* satisfies this condition. *DemoAnalyzer* is the graphical interface of the duel problem. One see it, if the *DuelStarwars* task is on. Figure 8.9 shows this.

Optimization results and the convergence line is in Figure 8.10.

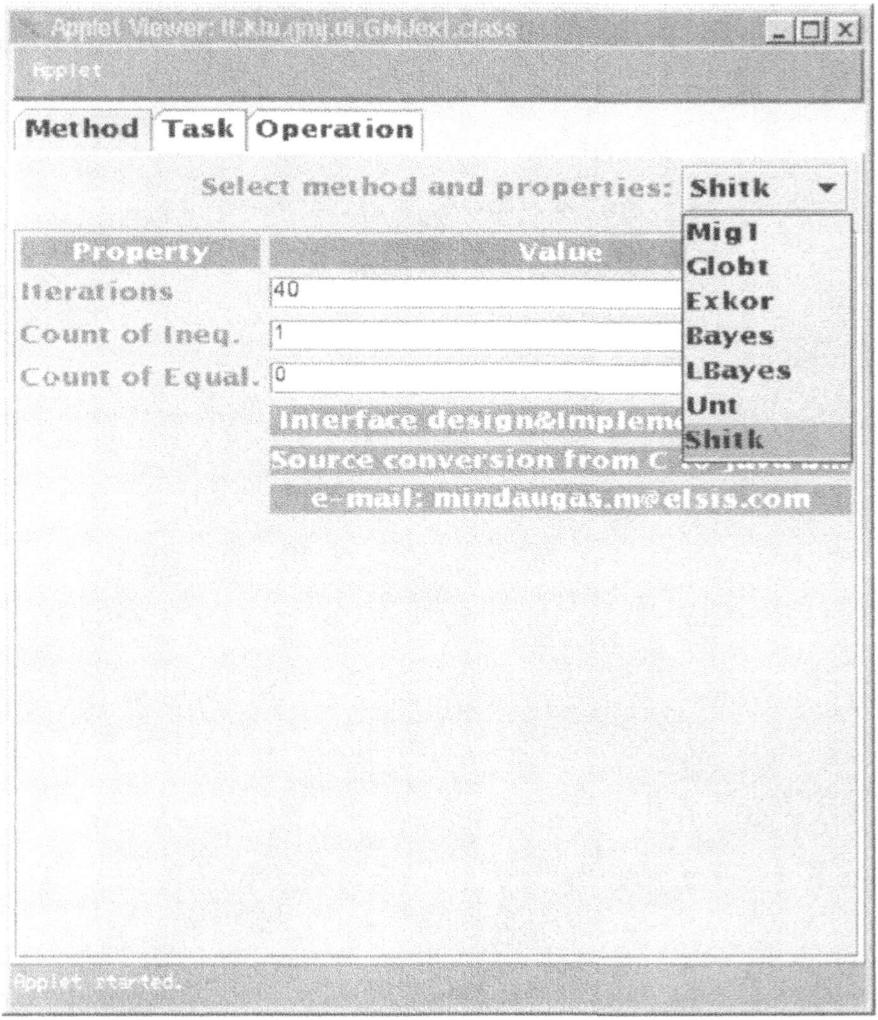

Figure 8.7. The method window of GMJ2.

In the 'Spectrum' window there is an additional option. The window shows how depends the distribution of $f(x)$ values on each variable x_i (see Figure 8.11).

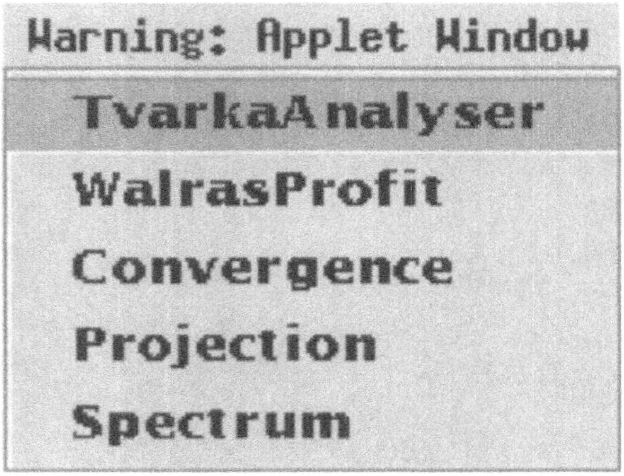

Figure 8.8. The analysis menu.

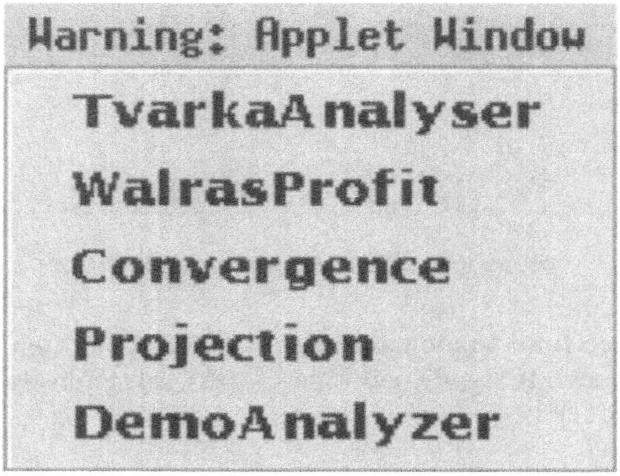

Figure 8.9. The analysis menu including the *DuelStarwars* task.

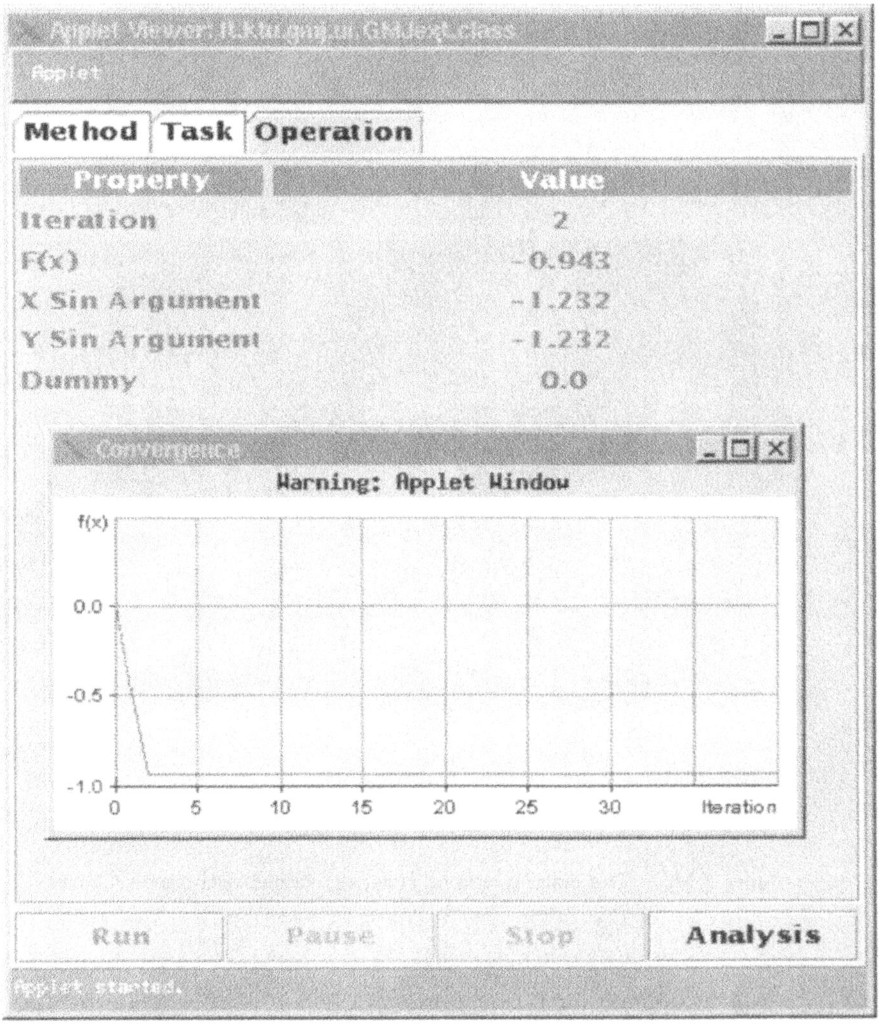

Figure 8.10. **Optimization results and the convergence line.**

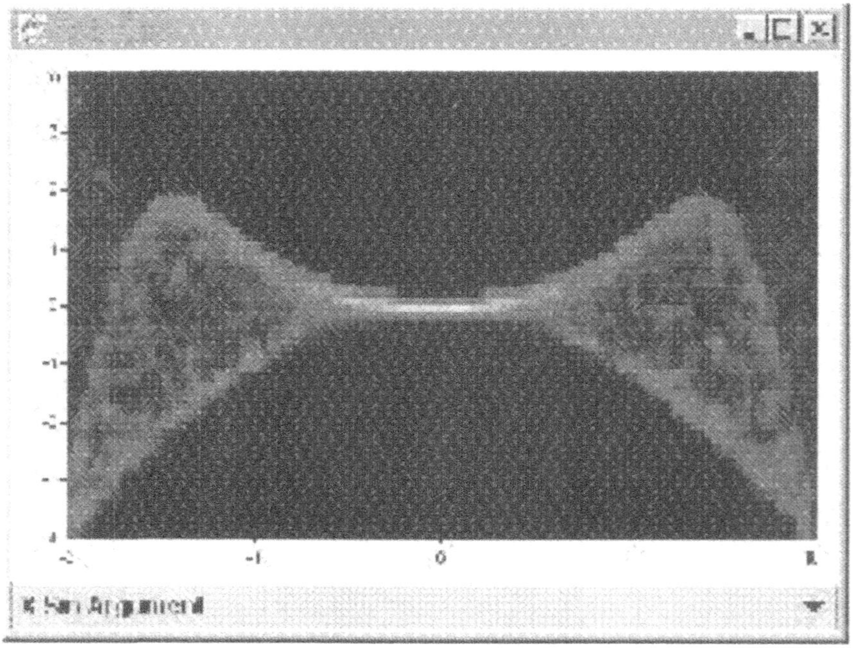

Figure 8.11. The distribution of $f(x_1, x_2)$ values with respect to x_1.

III

EXAMPLES OF MODELS

Chapter 9

COMPETITION MODEL WITH FIXED RESOURCE PRICES, NASH EQUILIBRIUM

1. OPTIMIZATION PROBLEMS IN MARKET MODELS

We consider two models of the theory of games and markets (Rosenmuller, 1981). The Nash equilibrium (Nash, 1950) is applied. The "market" is represented by a collection of independent servers. Each server tries to maximize its profit by setting optimal service prices and optimal server rates. The server rate is the average number of customers that would be served in nonstop operation. Thus, we can consider this rate as the server capacity.

The profits of individual servers are maximized assuming that their partners respect some agreement about the service prices and rates. We call this the "Contract-Vector." Service prices and rates obtained by maximizing individual profits of the servers transform the Contract-Vector into the "Fraud-Vector." The optimization problem is to search for such Contract-Vector that reduces the deviation of the Fraud-Vector from the Contract-Vector. That makes the fraud less relevant. The fraud is irrelevant and the Nash equilibrium is achieved, if this deviation is zero. Then servers cannot increase their profits by changing service prices and server rates agreed by the contract.

The quality of service is defined by the average time lost by customers while waiting for services. Formal expressions of other quality measures are difficult. A customer prefers the server with lesser total service cost.

The total cost includes the service price plus waiting losses. A customer goes away, if the total cost exceeds a certain critical level. A flow of customers is stochastic. Service times are stochastic, too. There is no known analytical solution for this model. The results are obtained by Monte-Carlo simulation.

The first market model considers a simple case. Here the server rate depends on a single resource that is freely obtainable at some market price. This model is the "Nash model," for short. In the second model, it is supposed that the server rates depend on several resources. Each server owns a fixed amount of single resource. Thus, the number of different resources is assumed to be equal to the number of servers. The resource limits are controlled using resource prices that are expressed as Lagrange multipliers[1]. This model is the "Walras model," for short. It is similar, but not identical, to the traditional Walras model (Walras, 1874).

In the Nash model, the servers rent their equipment at fixed price per unit of the server capacity. Therefore, in the Nash model we are looking for the equilibrium of server capacities and service prices. In the Walras model, the servers share each other resources. Therefore, here we are looking for the optimal prices charged for the shared resources. That is in addition to the search of equilibrium of server capacities and service prices.

Here the market models are to illustrate the possibilities and limitations of the optimization theory and numerical techniques in competitive environments. Both the models are developed as test functions for the Bayesian algorithms. However, these simple models may help to design more realistic ones describing the processes of competition better. Besides, the simplified competitive models are convenient tools for teaching the Operations Research and the Theory of Games and Markets.

2. NASH MODEL

Let us consider m servers providing the same service

$$u_i = u_i(x_1, y_1, ..., x_m, y_m) = a_i y_i - x_i, \; i = 1, ..., m, \tag{9.1}$$

[1] Defined by minimizing the Lagrange function corresponding to each server.

where u_i is the profit, y_i is the service price, a_i is the rate of customers, x_i is the running cost, and i is the server index. Assume that a server capacity w_i is an increasing function of the running cost x_i:

$$w_i = \phi_i(x_i). \tag{9.2}$$

A simple example of this function

$$w_i = k_i(1 - \exp(-k_{i0}x_i)). \tag{9.3}$$

Here k_i defines the maximal server rate and k_{i0} shows the efficiency of resource x_i. Then the total service cost

$$c_i = y_i + \gamma_i, \tag{9.4}$$

where γ_i is the average waiting cost. To simplify the expressions assume that the waiting cost is equal to an average waiting time. Suppose that arriving customers estimate the average waiting time as the relation of the number of waiting customers n_i to the server capacity w_i

$$\gamma_i = n_i/w_i \tag{9.5}$$

A customer goes to the server i, if

$$c_i \leq c_j,\ j = 1, ..., m,\ j \neq i,\ c_i \leq c_0. \tag{9.6}$$

A customer goes away, if

$$\min_i c_i > c_0, \tag{9.7}$$

where c_0 is the critical cost. The rate a of incoming consumers is fixed

$$a = \sum_{i=0}^{m} a_i, \tag{9.8}$$

where a_0 is the rate of lost customers. Conditions (9.6) and (9.7) separate the flow of incoming customers into $m+1$ flows. This makes the problem very difficult for analytical solution. The separated flow is not simple one, even in the case when the incoming flow is Poisson (Gnedenko and Kovalenko, 1987). Thus, we need the Monte Carlo simulation, to define average rates of customers a_i, $i = 0, 1, ..., m$, by conditions (9.6) (9.7), and average profits u_i, $i = 1, ..., m$ by expression (9.1).

3. SEARCH FOR NASH EQUILIBRIUM

First we fix the initial values, the "Contract-Vector" $z^0 = (x_i^0, y_i^0,\ i = 1, ..., m)$. The transformed values, the "Fraud-Vector" $z^1 = (x_i^1, y_i^1,\ i = 1, ..., m)$, is obtained by maximizing the profits of each server i. The maximization is performed under the assumption that all partners $j \neq i$ will honor the contract $(x_j^0, y_j^0,\ j = 1, ..., m,\ j \neq i)$

$$(x_i^1, y_i^1) = arg \max_{x_i, y_i} u_i(x_i, y_i, x_j^0, y_j^0,\ j = 1, ..., m,\ j \neq i),\ i = 1, ..., m. \quad (9.9)$$

Formally, condition (9.9) transforms the vector $z^n = (x_i^n, y_i^n,\ i = 1, ..., m) \in B \subset R^{2m},\ n = 0, 1, 2, ...$ into the vector z^{n+1}. To make expressions shorter denote this transformation by T

$$z^{n+1} = T(z^n),\ n = 0, 1, 2, ... \quad (9.10)$$

One may obtain the equilibrium at the fixed point z^n, where

$$z^n = T(z^n). \quad (9.11)$$

The fixed point z^n exists, if the feasible set B is convex and all the profit functions (9.1) are convex (Michael, 1976). We obtain the equilibrium directly by iterations (9.10), if the transformation T is contracting (Neuman and Morgenstern, 1953). If not, then we minimize the square deviation

$$\min_{z \in B} \| z - T(z) \|^2 . \quad (9.12)$$

The equilibrium is achieved, if the minimum (9.12) is zero. If the minimum (9.12) is positive then the equilibrium does not exist. That is a theoretical conclusion. In statistical modeling, some deviations are inevitable. Therefore, we assume that the equilibrium exists, if the minimum is not greater then modeling errors.

One can minimize deviation (9.12) by the usual stochastic approximation techniques (Ermoljev and Wets, 1988), if square deviation (9.12) is an uni-modal function of z. If not, then the Bayesian techniques of global stochastic optimization (Mockus, 1989a) should be used. The global stochastic optimization may outperform the local one in the uni-modal case, too. That happens, if the noise level is great because the Bayesian global stochastic optimization methods are less sensitive to large noise levels.

The Equilibrium is stable, if transformation (9.10) is locally contracting near a fixed point (9.11). If not, then some stabilizing conditions should be introduced. The transformation $T(z)$ is referred to as locally contracting, if there exists a constant $0 \leq \alpha < 1$ such that

$$\|T(z^1) - T(z^2)\| \leq \alpha \|z^1 - z^2\| \quad (9.13)$$

for all $z^1, z^2 \in Z_\epsilon$. Here Z_ϵ is a ϵ-vicinity defined by the "natural" deviations from the equilibrium.

3.1 EXISTENCE OF NASH EQUILIBRIUM

The first existence theorem is due to Nash (Nash, 1951) and dates back to 1951. Many generalizations appeared since then. Finding less and less restrictive sufficient conditions have been an active field of research (Forgo et al., 1999). The proofs of these conditions are based on the various fixed point theorems (Brouwer, 1912; Kakutani, 1941; Browder, 1968). Considering the examples of this book, we prefer simple existence conditions to the general ones. Testing the existence conditions, we express them in terms of the profit functions u instead of the operators T.

For example, the equilibrium exists, if the profit $u(z)$ is strictly convex function of all the components of its parameters $z \subset Z$, and Z is a convex set (Michael, 1976). In such cases, small changes of z components will not change the maximum points considerably.

The situation would be different in the non-strictly-convex cases. Here even very small change of some parameters z, may change the maximum point z^* of $u(z)$ considerably. For example, in linear cases this point may jump from minimal to maximal limits. In multi-modal cases the point z^* can jump from one local minimum to another one. These sharp changes violate the continuity of the transformation T. The continuity of T is needed in the Brouwer's fixed point theorem (Brouwer, 1912). In other theorems, such as Kakutani's (Kakutani, 1941) or Browder's (Browder, 1968), the fixed-point conditions are less restrictive. However, testing these conditions is not a trivial task.

4. STABLE COALITION, CORE OF GAMES

If $m > 2$, the formation and stability of a coalition $S \subset M$ are important issues.
Denote by $S = i$ a coalition of an individual server i.

Denote by $S = (i_1, i_2)$ a coalition of two servers i_1 and i_2.
Denote by $S = (i_1, ..., i_m) = M$ a coalition of all m servers.
Assume that service prices $y_i = y_i(S)$, service rates $x_i = x_i(S)$, and shares of profit $u_i = u_i(S)$ are determined by the coalition agreement for all the coalition members $i \in S$.
Denote by $|S|$ the number of servers in the coalition S and by $u(S) = \sum_{i \in S} u_i$. Suppose that all the remaining servers form the opposite coalition $M \backslash S$. Assume that the profits u_i corresponds to equilibrium service prices and rates $y(S), x(S), y(S), y(M \setminus S)$. This way we consider the m-server system as a set of two-server systems, where the first server is coalition S and the second one is coalition $M \setminus S$. A coalition $S^* \subset M$ is stable if there is no dominant coalition $S \subset M$ such that

$$u_i(S) > u_i(S^*), \; for \; all \; i \in S \tag{9.14}$$

The definition of the stable coalition is related to the definition of a game core $\mathbf{C}$ (Rosenmuller, 1981). It is well known that if

$$u(M) > \sum_{i \in M} u_i(M), \tag{9.15}$$

$$u(S) + u(M \setminus S) = u(m) \tag{9.16}$$

then the game core is empty $\mathbf{C} = \emptyset$ and there is no stable coalition S.

Inequality 9.15 means that a game is essential. Equality 9.16 defines a constant-sum game (Owen, 1968). The server system is clearly not a constant-sum game. Thus one of two "non-existence" conditions (9.16) is not satisfied meaning that a stable coalition S^* may exist. If a stable coalition S^* exists, one may determine it by testing 9.14 for all pairs of coalitions. The game core C is empty, if for any $S' \subset M$ there is a dominant coalition S.

4.1 EXAMPLE OF SERVER COALITION

Consider a case of three servers $m = 3$.
Denote by $S_1 = i$ the coalition of an individual server i.
Denote by $S_2 = (i_1, i_2)$ the coalition of two servers i_1 and i_2.
Denote by $S_3 = (i_1, i_2, i_3)$ the monopolistic coalition of all three servers.
Subscripts $k = 1, 2, 3$ denote numbers of servers in the coalition S_k.
Consider for simplicity the symmetric case. Here profit functions u and control parameters x, y depend only on the number of servers in the coalition but not on their "names."

The symmetric case implicitly assumes equality of the control parameters and equal sharing of the coalition profit, meaning that in the case of S_3

$$u_i(S_3) = (u_i + u_j + u_l)/3, \tag{9.17}$$

$$x_i(S_3) = x_j = x_l, \; y_i(S_3) = y_i = y_j = y_l. \tag{9.18}$$

In the case of S_2

$$u_i(S_2) = (u_i + u_j)/2, \quad x_i(S_2) = x_i = x_j, \; y_i(S_2) = y_i = y_j. \tag{9.19}$$

In the case of individual servers S_1

$$u_i(S_1) = u_i, \; x_i(S_1) = x_i, \; y_i(S_1) = y_i. \tag{9.20}$$

The monopolistic coalition S_3 is stable if $u_i(S_3) \geq u_i(S_2)$ and $u_i(S_3) \geq u_i(S_1)$.
The coalition S_2 is stable if $u_i(S_2) \geq u_i(S_3)$ and $u_i(S_2) \geq u_i(S_1)$.
The coalition S_1 is stable if $u_i(S_1) \geq u_i(S_3)$ and $u_i(S_1) \geq u_i(S_2)$.
There is no stable coalition if no of these condition are satisfied.

Chapter 10

COMPETITION MODEL WITH FREE RESOURCE PRICES, WALRAS EQUILIBRIUM

1. WALRAS MODEL

In the model called by the Nash name, the cost of a service capacity unit is supposed to be fixed. Each individual server i controls the capacity x_i and the price y_i charged for services. Therefore, the Nash model illustrates the formation of competitive service prices, if the resource prices are fixed by some large external market.

More complicated models, called by the Walras name, describe the formation of the competitive resource prices, too. In the Walras model, the capacity w_i of servers $i = 1, ..., m$ depends on the resource vector $x_i = (x_{ij},\ i, j = 1, ..., m)$ defining the consumption of different resources. Each server i owns a single resource b_i and charges a price p_i for a unit of this resource to partner-servers and to himself.

A server i controls the resource vector $x_i = (x_{ij},\ j = 1, ..., m)$. The component x_{ij} denotes the amount of resource b_j used by the server i. The server i also controls the price y_i of services, as in the Nash model. Therefore we shall describe the Walras model in the terms similar to those of the Nash model (see expression (9.1)).

In the Walras model, we introduce the notion of a "penalty" v_i for the missing resource b_i. The penalty v_i is

$$v_i = -(1 + \alpha) p_i (b_i - \sum_{k=1}^{m} x_{ki}),\ i = 1, ..., m \tag{10.1}$$

Here x_{ki} denotes the amount of resource b_i used by the server k. That means that the server i pays penalty v_i if the system is short of the resource b_i. Note, that the penalty turns into a premium if there is excess of this resource.

In the two-server case from expression (10.1)

$$v_1 = -(1+\alpha)p_1(b_1 - x_{11} - x_{21}) \tag{10.2}$$

and

$$v_2 = -(1+\alpha)p_2(b_2 - x_{22} - x_{12}) \tag{10.3}$$

Using the multiplier $1 + \alpha$ one can define different penalty scales. For example, this expression

$$\alpha = \begin{cases} \alpha_1, & \text{if } v_i \geq 0 \\ \alpha_2, & \text{if } v_i < 0, \end{cases} \tag{10.4}$$

where $\alpha_1 > \alpha_2$ means, the server pays more for the missing resource comparing to what it gets for the excess one. Therefore, in the Walras model software α_1 is called as "credit rate" and α_2 is called as"deposits yield."

The parameter $\epsilon \geq 0$ introduces a small non-linear component of resource prices. The price of resource I is expressed as the product $p_i(1 + \epsilon p_i)$. The parameter ϵ has no economical meaning. It helps to stabilize the computing process, sometimes.

For unit scales, where $\alpha_1 = \alpha_2 = \epsilon = 0$, the prices p_i are considered as Lagrange multipliers. Therefore, we assume that $\alpha_i = 0$ and $\epsilon = 0$, as usual. We define the equilibrium resource prices p_i by minimizing the Lagrange function in the framework of the traditional method of Lagrange multipliers (Horst et al., 1995).

Assume that the server capacity w_i is an increasing function of the resource vector $x_i = (x_{ij}, j = 1, ..., m)$.

$$w_i = \phi_i(x_i). \tag{10.5}$$

A simple example of this function

$$w_i = k_i \prod_{j=1}^{m} (1 - \exp(-k_{ij}x_{ij})). \tag{10.6}$$

The resource component x_{ij} denotes the amount of resource b_j used by server i. The coefficient k_{ij} shows how useful is the resource b_j for the server i. The coefficient k_i defines the capacity limit when $x_{ij} \to \infty$. Then the profit of the i-th server:

$$u_i = u_i(x_j, y_j, p_j, \ j = 1, ..., m) = \\ a_i y_i - v_i - \sum_{j \neq i} p_j x_{ij}. \tag{10.7}$$

Here i is the server index. a_i is the rate of customers. y_i is the service price. v_i is the penalty for overusing the resource b_i. $x_j = (x_{jk}, \ k = 1, ..., m)$ is the resource vector determining the capacity w_j of server j. x_{jk} denotes the amount of resource b_k used by the server j. Note, he profit u_i of each individual server i depends on the parameters x_j, Y_j, p_j of all m servers $j = 1, ..., m$. In the Lagrangian case where $\alpha_i = 0$ and $\epsilon = 0$, the profit

$$u_i = a_i y_i - p_i(\sum_{k=1}^{m} x_{ki} - b_i) - \sum_{j \neq i} p_j x_{ij}. \tag{10.8}$$

Here the first component $a_i y_i$ defines the income of a server i. The second component $p_i(\sum_{k=1}^{m} x_{ki} - b_i)$ defines the penalty for over-using the resource b_i. The third component $\sum_{j \neq i} p_j x_{ij}$ shows the charge for resources obtained from other servers. In Lagrangian two-server cases

$$u_1 = u_1(x_{11}, x_{12}, y_1, p_1, x_{22}, x_{21}, y_2, p_2) = \\ a_1 y_1 - p_1(x_{11} + x_{21} - b_1) - p_2 x_{12} \tag{10.9}$$

and

$$u_2 = u_2(x_{22}, x_{21}, y_2, p_2, x_{11}, x_{12}, y_1, p_1) = \\ a_2 y_2 - p_2(x_{22} + x_{12} - b_2) - p_1 x_{21} \tag{10.10}$$

Assuming the lower and upper limits
$a_{p_i}, a_{x_{ij}}, a_{y_i}, b_{p_i}, b_{x_{ij}}, b_{y_i}$ $i, j = 1, ..., m$ we obtain the inequalities

$$a_{p_i} \leq p_i \leq b_{p_i}, \ a_{x_{ij}} \leq x_{ij} \leq b_{x_{ij}}, \\ a_{y_i} \leq y_i \leq b_{y_i} \tag{10.11}$$

Here the service cost, the waiting cost, and the customer behavior are similar to those of the Nash model. Namely, the service cost

$$c_i = y_i + \gamma_i, \tag{10.12}$$

where γ_i is waiting cost[1] at the server i (see expression (9.5)) . A customer goes to the server i, if

$$c_i < c_j, \ j = 1, ..., m, \ j \neq i, \ c_i \leq c_0. \tag{10.13}$$

A customer goes away, if

$$\min_i c_i > c_0, \tag{10.14}$$

where c_0 is the critical cost. The rate a of incoming consumers flow is fixed:

$$a = \sum_{i=0}^{m} a_i, \tag{10.15}$$

where a_0 is the rate of lost customers.

2. SEARCH FOR WALRAS EQUILIBRIUM

We fix a contract-vector $(x_i^0, y_i^0, p_i^0, \ i = 1, ..., m)$. Then the fraud-vector $(x_i^1, y_i^1, p_i^1, \ i = 1, ..., m)$ is obtained by maximizing the profits of each server i and by assuming that all the partners $j \neq i$ will respect the contract $(x_i^0, y_i^0, p_i^0, \ i = 1, ..., m)$

$$(x_i^1, y_i^1, p_i^1,) = \\ arg \min_{p_i} \max_{x_i, y_i} u_i(x_i, y_i, p_i, \ x_j^0, y_j^0, p_j^0, \ j = 1, ..., m, \ j \neq i). \tag{10.16}$$

Here the profit function $u_i(x_i, y_i, p_i, \ x_j^0, y_j^0, p_j^0, \ j = 1, ..., m, \ j \neq i)$ is defined by expression (10.8). Lagrange multipliers p_i are obtained by the minimization of the Lagrangian (10.8) under rectangular constraints (10.11). In the two-server case

$$(x_{11}^1, x_{12}^1, y_1^1, p_1^1,) = \\ arg \min_{p_1} \max_{x_{11}, x_{12}, y_1} u_1(x_{11}, x_{12}, y_1, p_1, x_{22}^0, x_{21}^0, y_2^0, p_2^0), \tag{10.17}$$

and

$$(x_{22}^1, x_{21}^1, y_2^1, p_2^1,) = \\ arg \min_{p_2} \max_{x_{22}, x_{21}, y_2} u_2(x_{22}, x_{21}, y_2, p_2, x_{11}^0, x_{12}^0, y_1^0, p_1^0). \tag{10.18}$$

[1]It is. assumed for simplicity that the waiting cost is equal to an average waiting time.

Condition (10.16) transforms vectors z^n, $n = 0, 1, 2, ...$ into vectors z^{n+1}, where $z^n = (x^n, y^n, p^n)$, $x^n = (x_1^n, ..., x_m^n)$, $y^n = (y_1^n, ..., y_m^n)$, and $p^n = (p_1^n, ..., p_m^n)$. Denote this transformation by T

$$z^{n+1} = T(z^n),\ n = 0, 1, 2, ... \tag{10.19}$$

Here the vector $z = (x_i, y_i, p_i, v_i\ i = 1, ..., m) \in B \subset R^{m^2+2m}$. We reach the equilibrium[2] at the fixed point z^n, where

$$z^n = T(z^n). \tag{10.20}$$

We may obtain the equilibrium directly by simple iterations (10.19), if the transformation T is contracting (Neuman and Morgenstern, 1953). There are more sophisticated and efficient iterative procedures (Herings, 1994).

If the equilibrium exists but the transformation T is not contracting then we minimize the square deviation

$$\min_{z \in B} \| z - T(z) \|^2 . \tag{10.21}$$

The equilibrium is achieved, if the minimum (10.21) is zero. The feature of the Walras model (Walras, 1874) is that there are no excess demand or surplus of resources x_i, if the resource prices p_i meet the equilibrium condition. We use this as second test of equilibrium, besides zero of minimum conditions (10.21).

$$\sum_{i=1}^{m} (|\sum_{j=1}^{m} x_{ij} - b_i|) = 0. \tag{10.22}$$

The constraints (10.11) limits the Walras model. Therefore, setting these constraints one should resolve the following contradiction. Wider limits means more computing for optimization but less restriction for the Walras model, and vice versa. A way to get around this contradiction is to start with narrow bounds (10.11). One widens them later. Some of these bounds obviously restrict the profit maximum (see for example Figure 10.1). The figure shows that the lower limit $x11 \geq 1$ restricts the maximal profit.

The equilibrium tests (10.21) and (10.22) fail, if there is no equilibrium. That means the problem is ill-defined and the Walras model

[2] If the equilibrium exists.

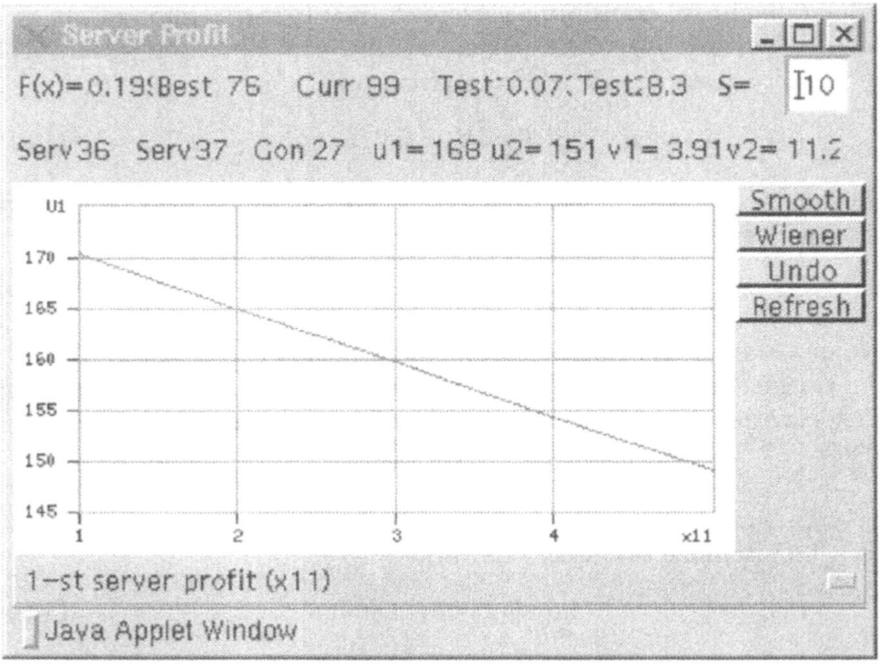

Figure 10.1. The relation of the profit u_1 on the resource x_{11}.

should be corrected. In such a case, additional testing of the existence conditions is needed. It is well known, that the equilibrium exists, if the profit u is a convex function of parameters y, x, p (Michael, 1976). Therefore, the "non-convexity" of this function can be a reason of the failure to obtain the equilibrium.

Another reason of the failure might be insufficient optimization. Here the equilibrium can be reached by increasing the number of iterations (if the optimization method converges).

3. MONTE-CARLO SIMULATION

3.1 SEARCH FOR EQUILIBRIUM

The analytical solution of both the described market models is not practical. Therefore, we briefly consider the basic steps of an algorithm of the statistical simulation using Monte-Carlo techniques. The algorithm implements two basic tasks:

- generates the next event time t,
- updates the state of queuing system defined by the vector of waiting customers $n(t) = (n_i(t),\ i = 1, ..., m)$,
- updates the vector $h(t) = (h_i(t),\ i = 1, ..., m)$ of the service cost including the money charged and the time lost.

There are $2m + 2$ types of event times t:

- the time t when a customer arrives into the system,
- the time t when a customer arrives into the i-th server, $i = 1, ..., m$,
- the time t when a customer departs from the i-th server,
- the time t when a customer abandons the service (departs from the system without being served).

Here $i = 1, ..., m$. The system state is updated at each event time t. Two vectors define the system state:

- a vector $n = n(t)$ with m components $n = (n_1, ..., n_m)$, where $n_i = n_i(t)$ shows the number of customers waiting for the service of the i-th server,
- a vector $h = h(t)$ with m components $h = (h_1, ..., h_m)$, where $h_i(t) = y_i + \gamma_i$, $\gamma_i = n_i(t)/w_i$ shows the total customer expenses, y_i is money charged for the service, and γ_i is the time lost waiting for the service of the i-th server[3].

There are no state changes between events. The basic steps of the Monte Carlo algorithm:

1. fix the zero event time $t = t^0 = 0$ when the first customer arrives,

[3]For simplicity, it is assumed that "time-is-money" and that a unit of time cost a unit of money, in the real life cases the corresponding coefficients should be included

2. define the zero state vector n^0 by the condition: $n_i^0 = 0,\ i = 1, ..., m$ and the zero state vector h^0 by the condition : $h_i^0 = y_i = 0,\ i = 1, ..., m$
because there are no customers waiting for service yet,

3. define the next arrival into the system by the expression

$$\tau_a = -1/a\ ln(1 - \eta) \tag{10.23}$$

where η is a random number uniformly distributed in the interval [0,1],

4. chose the best server i^0 for the first customer by the condition $i^0 = arg \min_{i=0,...,m} h_i$ where $h_i = y_i$, because $\gamma_i = 0,\ i = 1, ..., m$ since there are no customers waiting yet, $i^0 = 0$ means that the customer abandons the service,

5. define the time of event when the first customer will be served by the server i^0 using the expression

$$\tau_{i^0} = -1/x_{i^0}\ ln(1 - \eta), \tag{10.24}$$

6. define the next event t^1 by comparing the arrival time τ_a and the service time τ_{i^0} :
if $\tau_a < \tau_{i^0}$ then $t^1 = \tau_a$,
if $\tau_a > \tau_{i^0}$ then $t^1 = \tau_{i^0}$,

7. define the system state at the next event t^1:
if $t^1 = \tau_a$ then $n_{i^0} = 1$ and $n_i = 0,\ i = 1, ..., m,\ i \neq i^0$,
consequently $h_{i^0} = y_{i^0} + 1/w_{i^0}$ and $h_i = y_i,\ i = 1, ..., m,\ i \neq i^0$,
if $t^1 = \tau_{i^0}$ then $n_i = 0,\ i = 1, ..., m$, and $h_i = y_i,\ i = 1, ..., m,\ i \neq i^0$.

Definition of later events and system states is longer but the main idea remains the same.

The algorithm can be directly adapted to the Monte-Carlo simulation of the Nash model with two servers, too. For example, that can be done this way:
- set to unit both the resource charges $p_i = 1,\ i = 1, 2$,
- set to zero resources exchanges $x_{21} = x_{12} = 0$,
- assume that $x_{11} = x_1,\ x_{22} = x_2$, where variables $x_1,\ x_2$ are from expression (9.1).

If the number of severs $m > 2$ then some modification of the described algorithm is needed.

3.2 TESTING EQUILIBRIUM CONDITIONS

In the Monte-Carlo simulation, equilibrium tests (10.21) and (10.22) should be relaxed by accepting some simulation error ϵ:

$$\min_{z \in B} \| z - T(z) \|^2 \leq \epsilon, \tag{10.25}$$

and

$$\sum_{i=1}^{m} (|\sum_{j=1}^{m} x_{ij} - b_i|) \leq \epsilon. \tag{10.26}$$

Testing the convexity of profit functions, some smoothing is desirable. The smoothing eliminates the random deviations due to Monte-Carlo simulation. Both the convolution and the Wiener filters are applied for smoothing the profit functions (possibly multi-modal). The convolution filter defines the function at some fixed point as an average of values in the neighborhood of this point. The more sophisticated Wiener filter is implemented, too.

3.3 WIENER FILTER

If the objective function $f(x)$ is defined by Monte Carlo simulation, some noise is present. That means that one observes the sum

$$\phi(x) = f(x) + \xi, \tag{10.27}$$

where ξ is a random number called the noise.

If finding the optimum of a convex function $f(x)$ is the only goal, we can apply some stochastic optimization algorithms (Ermoljev and Wets, 1988). These algorithms converge to the optimum of $f(x)$ by filtering the noise during the optimization process.

To test properties of $f(x)$, such as convexity, uni-modality e.t.c., we need specific smoothing algorithms that eliminate false local optima. In one-dimensional cases, a convenient smoothing function is the conditional expectation of the Wiener process with noise (Kushner, 1964b; Senkiene, 1980). It is assumed that the optimization parameter $x \in [0, 1]$, the Wiener parameter is a unit, and the noise ξ is Gaussian with zero mean and variance S_i at the points $x^i \in [0, 1]$, $i = 1, ..., n$.

Then the conditional expectation $\mu_k = \mu_n(x^k)$ of the objective function $y_k = f(x^k)$ at some fixed point x^k with respect to the observations results $y_i = \phi(x^i),\ i = 1, ..., n$, can be expressed this way

$$\mu_k = \frac{\sum_{i=1}^{k} b_i y_i + \frac{b_k}{c_k}\sum_{i=k+1}^{n} c_i y_i}{\sum_{i=1}^{k} b_i + \frac{b_k}{c_k}\sum_{i=k+1}^{n} c_i}. \tag{10.28}$$

Here

$$b_1 = 1,\ B_1 = 1, \tag{10.29}$$

$$b_2 = \frac{S_1^2 + r_{1,2}}{S_2^2},\ B_2 = B_1 + b_2, \tag{10.30}$$

....

$$b_k = \frac{S_{k-1}^2 b_{k-1} + r_{k,k-1}B_{k-1}}{S_k^2},\ B_k = B_{k-1} + b_k, \tag{10.31}$$

$$c_n = 0, \tag{10.32}$$

$$c_k = \frac{S_{k+1}^2 c_{k+1} + r_{k,k+1}C_{k+1}}{S_k^2}, C_{k+1} = \sum_{i=k+1}^{n} c_i, \tag{10.33}$$

$$r_{i,j} = |x^i - x^j|. \tag{10.34}$$

It is convenient to assume that

$$S_i = S,\ i = 1, ..., n, \tag{10.35}$$

where S can be considered as a smoothing parameter.

If $S = 0$ then no smoothing occurs. The smoothing function (the conditional expectation) is the piece-wise line connecting the observed points (see Figure 1.1).
If S is large then one obtains a horizontal line corresponding to the average value of observed values y_i. That means a sort of "total smoothing." In modeling the yield of differential amplifiers, the best smoothing was achieved at $S = 10$ (Mockus et al., 1997).

Figure 10.2 shows how the first server profit u_1 depends on the price p_1 charged for its resources. There are two samples of the same relation. They show the differences between two samples of random arrival times of fifty customers.
The buttons 'smooth' and 'wiener' on the right side of Figure 10.9 are for switching on these filters. The button 'smooth' is for the convolution filter. The button 'wiener' is for the Wiener filter.
The field denoted by 'S' at the top right corner, defines the smoothing

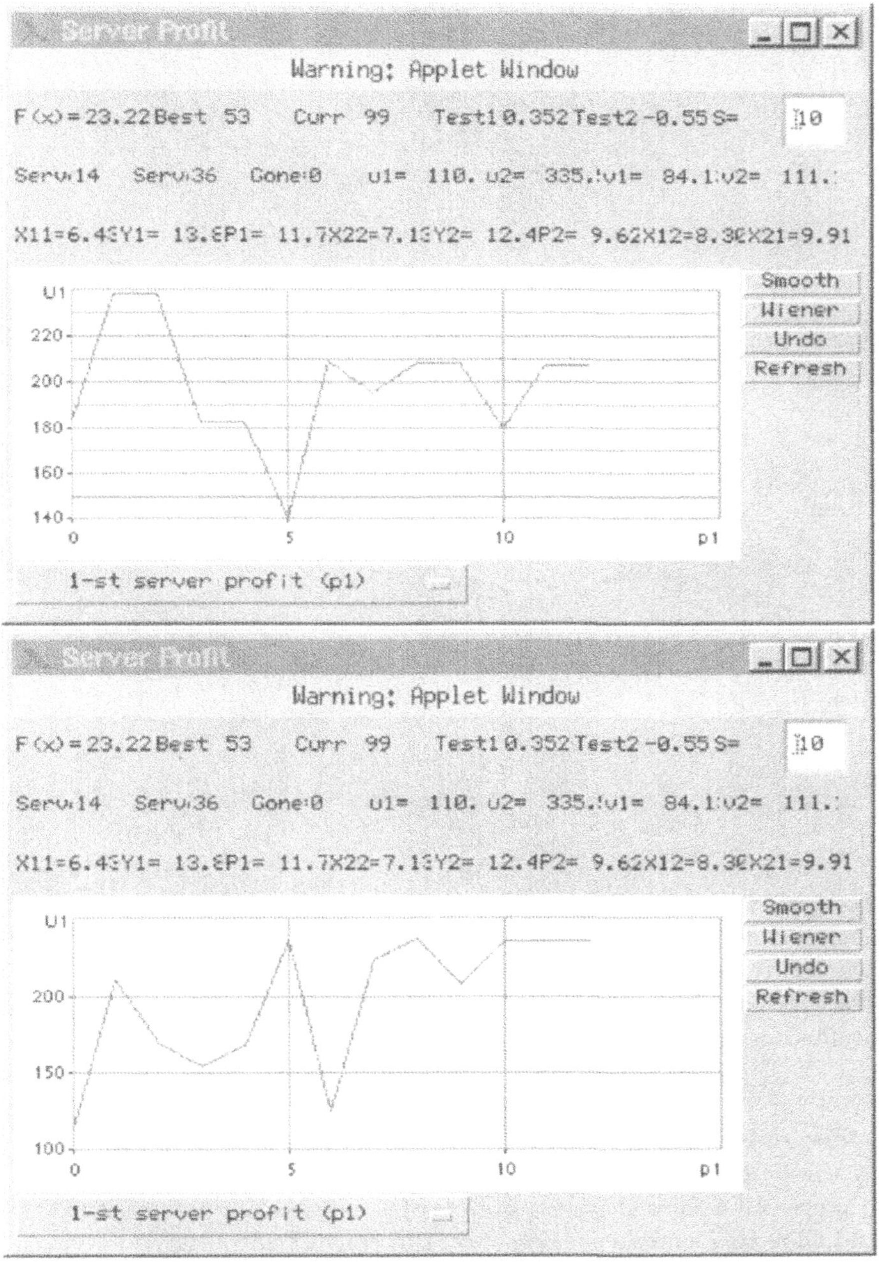

Figure 10.2. The relation of the profit u_1 on the resource price p_1 (top), the "refreshed" relation of the profit u_1 on the resource price p_1 (bottom).

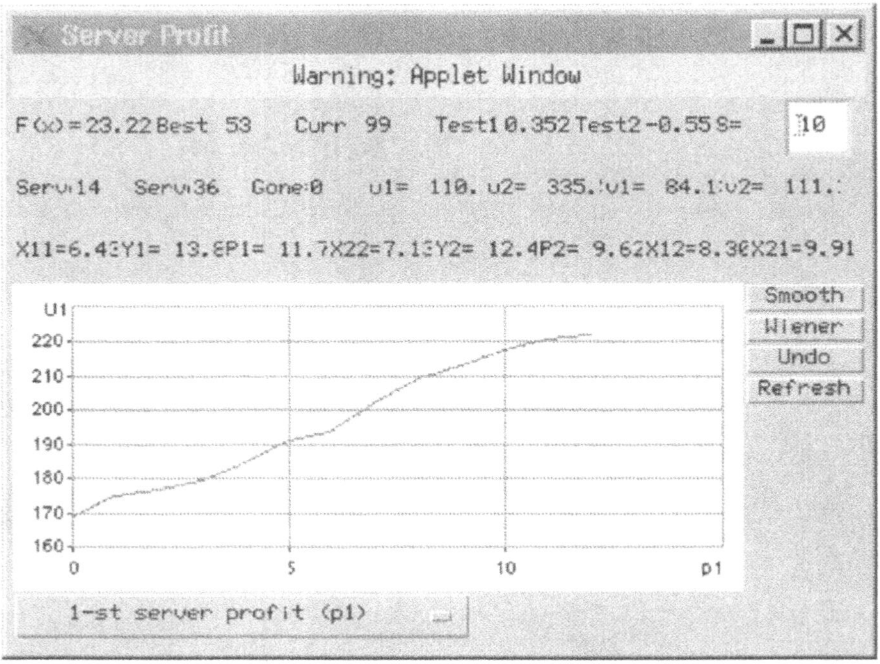

Figure 10.3. The relation of the profit u_1 on the resource x_{22} smoothed by the Wiener filter.

parameter S of the Wiener filter (see expression (10.35).
One can increase the level of smoothing by pressing the 'smooth' or 'wiener' buttons repeatedly.
The 'refresh' button repeats the Monte-Carlo simulation of the same profit function.

Figure 10.3 shows the graph shown in Figure 10.2 (bottom) smoothed by the Wiener filter.
Figures 10.4 show the third sample of the 'refreshed' unsmoothed graph and the same sample smoothed by the convolution filter.

The underlying profit function is the same in all the Figures, from 10.2 up to 10.4. The repeated simulation defines different graphs because of the simulation errors. After smoothing, the level of these errors is lower.

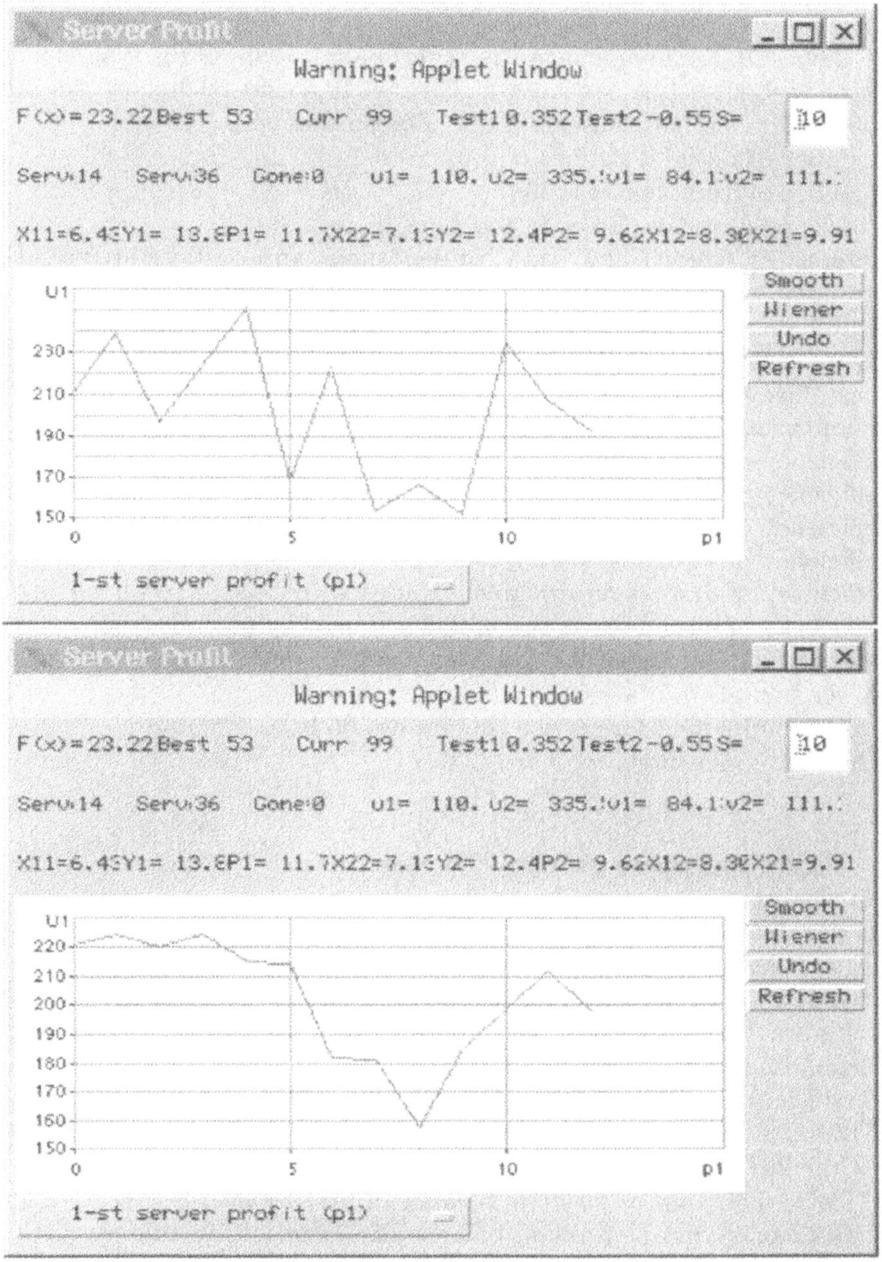

Figure 10.4. **The third sample of the relation of the profit u_1 on the resource p_1 (top), the relation of the profit u_1 on the price p_1, smoothed by the convolution filter (bottom).**

4. SOFTWARE EXAMPLE

Java JDK1.1.3 implementation (Perlibakas, 1999) of the Walras model is on web-sites (see section 4.). Figures 10.5 and 10.6 show the input page. The method 'Bayes1' is set. The parameters:
number of initial iterations lt is 5,
number of iterations it is 30,
stocks of server resources b_1 and b_2 is 16
interest of "credit" and "yeld" of deposit α_1 and α_2 both are zero (Lagrangian case)
stabilizing non-linearity ϵ is 0.01
"run-away" threshold $C0$ is 20
customer rate A is 50,
number of time units M is 1
efficiency of first server z_{01} is 8
efficiency of first server resource when used by first server z_{11} is 0.2
efficiency of second server resource when used by first server z_{21} is 0.2
efficiency of second server z_{02} is 8
efficiency of first server resource when used by second server z_{12} is 0.2
efficiency of second server resource when used by second server z_{22} is 0.2
steps of all parameter changes $Step$ are 1.0,
lower bounds min are 0.0,
upper limits max for service charges are 36.0,
the rest upper limits max are 12.0.

Figure 10.7 shows the results of optimization.
In this figure:
it denotes the "best' iteration
$F(x)$ means the minimal deviation from the equilibrium point,
$x_{11}, y_1, p_1, x_{22}, y_2, p_2, x_{12}, x_{21}$ are the optimal values of the parameters.

Figure 10.8 shows how the profits of first and second server change depending on the iteration number it.
The Figure provides some additional information, too.
"Best 20"denotes the best iteration.
"Curr 29" denotes the last iteration, both counts from zero.
"Test1 -0.50" means negative balance of first server resource.
"Test2 0.59" means positive balance of second server resource (see expression (10.22)).
"Serv 42" means that the first server served 42 customers.
"Serv 8" means that the second one served only 8.

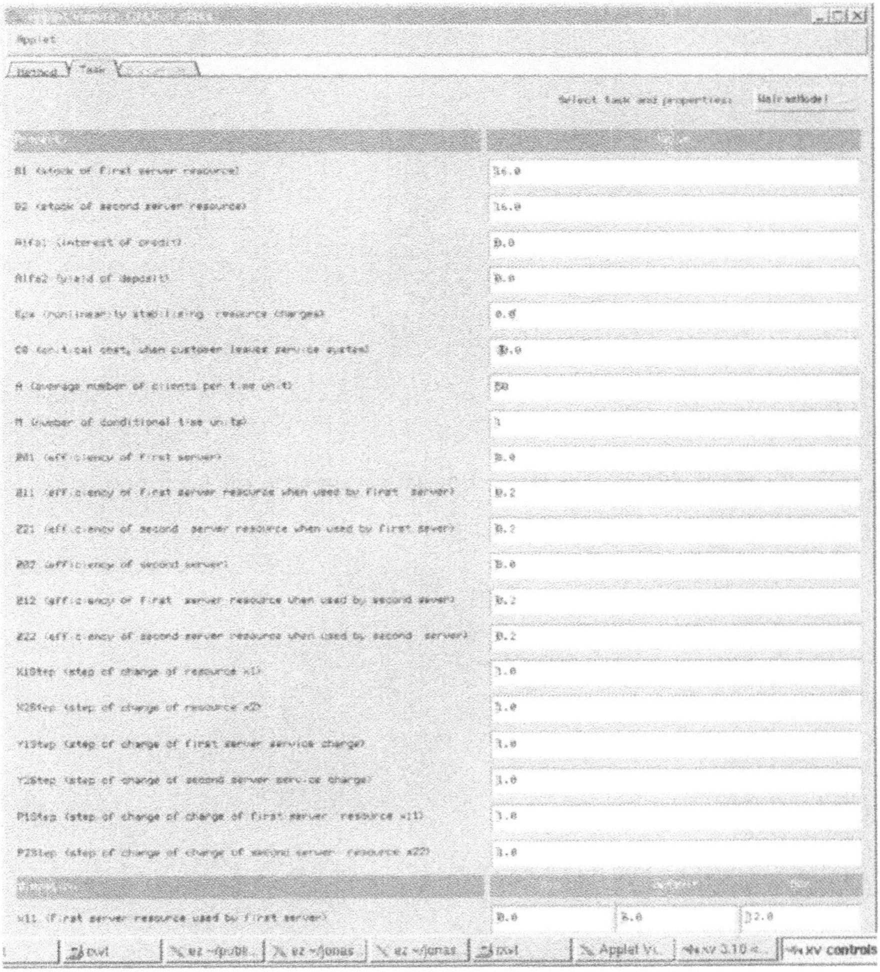

Figure 10.5. The table of initial parameters, the upper part.

"Gone 0" means that no customer was lost'.
"v1=99.2" and "v2=84.6" show the "penalty" functions (see expression (10.1)).
The rest numbers just repeat the optimal parameters shown in the output table (see Figure 10.7). Figure 10.9 shows how the profits of first and second server depend on the auxiliary resources x_{11} and x_{22}
Figure 10.10 show how the profits of first and second server depend on the resources obtained from their partners x_{12} and x_{21}.
Figure 10.11 show how the profits of first and second server depend on

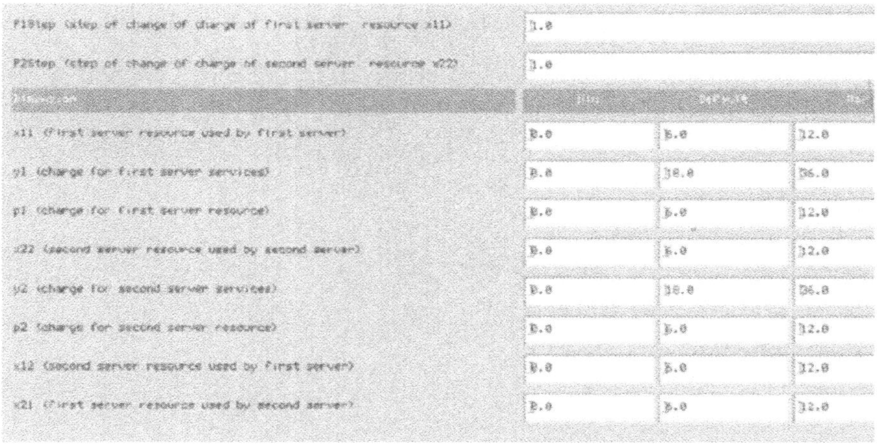

F1Step (step of change of charge of first server resource x11)	1.0		
F2Step (step of change of charge of second server resource x22)	1.0		
x11 (first server resource used by first server)	0.0	6.0	12.0
y1 (charge for first server services)	0.0	18.0	36.0
p1 (charge for first server resource)	0.0	6.0	12.0
x22 (second server resource used by second server)	0.0	6.0	12.0
y2 (charge for second server services)	0.0	18.0	36.0
p2 (charge for second server resource)	0.0	6.0	12.0
x12 (second server resource used by first server)	0.0	6.0	12.0
x21 (first server resource used by second server)	0.0	6.0	12.0

Figure 10.6. The table of initial parameters, the lower part.

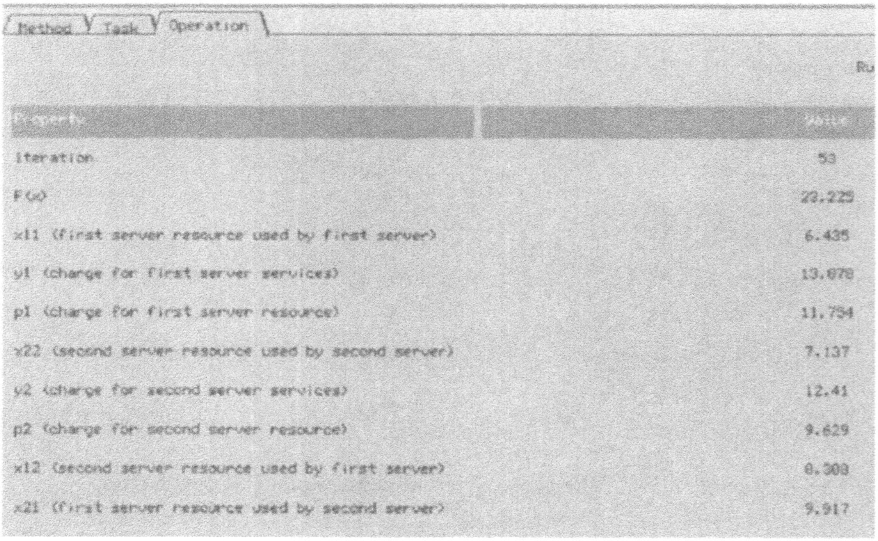

Property	Value
Iteration	53
F(x)	23.225
x11 (first server resource used by first server)	6.435
y1 (charge for first server services)	13.078
p1 (charge for first server resource)	11.754
x22 (second server resource used by second server)	7.137
y2 (charge for second server services)	12.41
p2 (charge for second server resource)	9.629
x12 (second server resource used by first server)	8.308
x21 (first server resource used by second server)	9.917

Figure 10.7. Optimization results using the method Bayes1.

the service charges y_2 and y_2.

The figures show that optimization is approaching the equilibrium. However, to obtain better accuracy more computing power is needed. Simulating fifty customers, one iteration needs five minutes of Pentium 300 processor time.

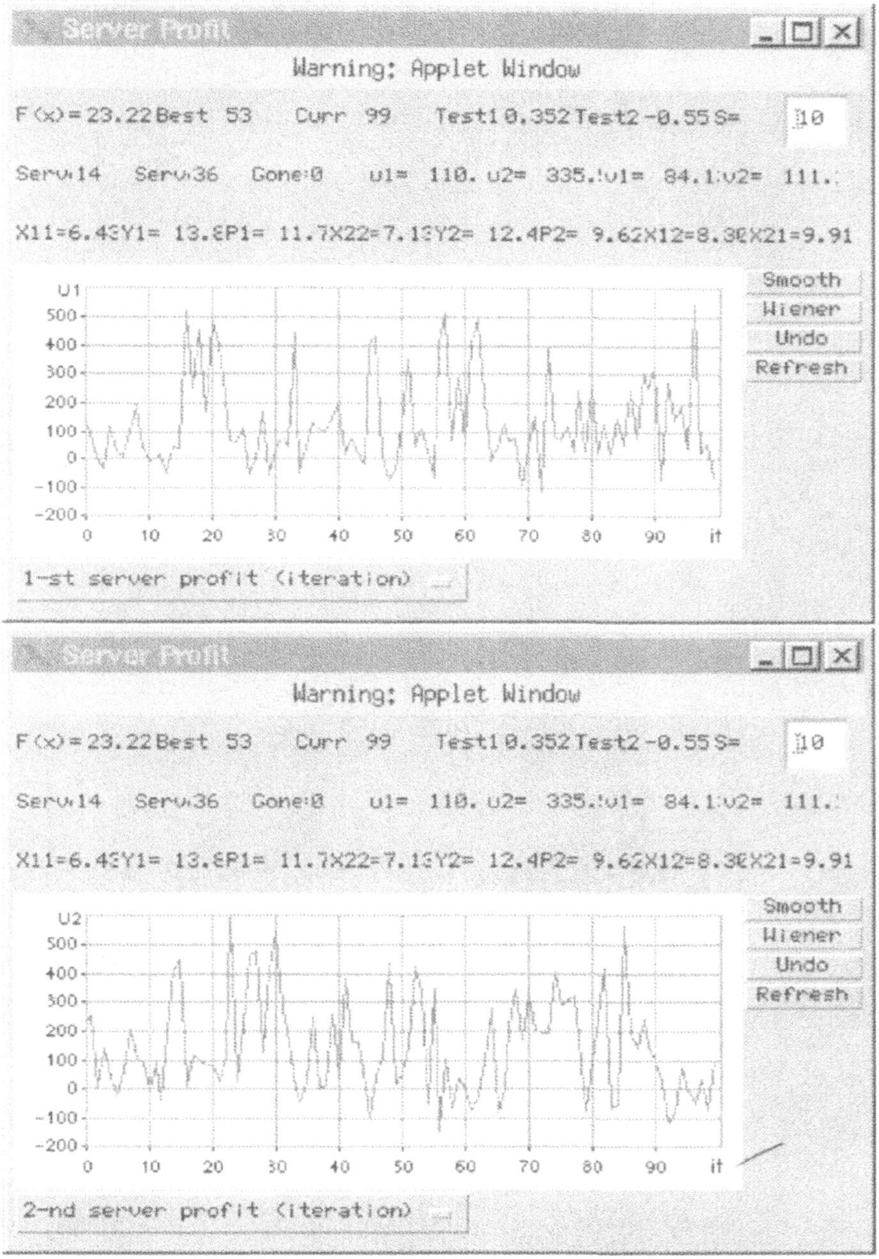

Figure 10.8. Relation of the profit u_1 on the iteration number *it* (top), Relation of the profit u_2 on the iteration number *it* (bottom).

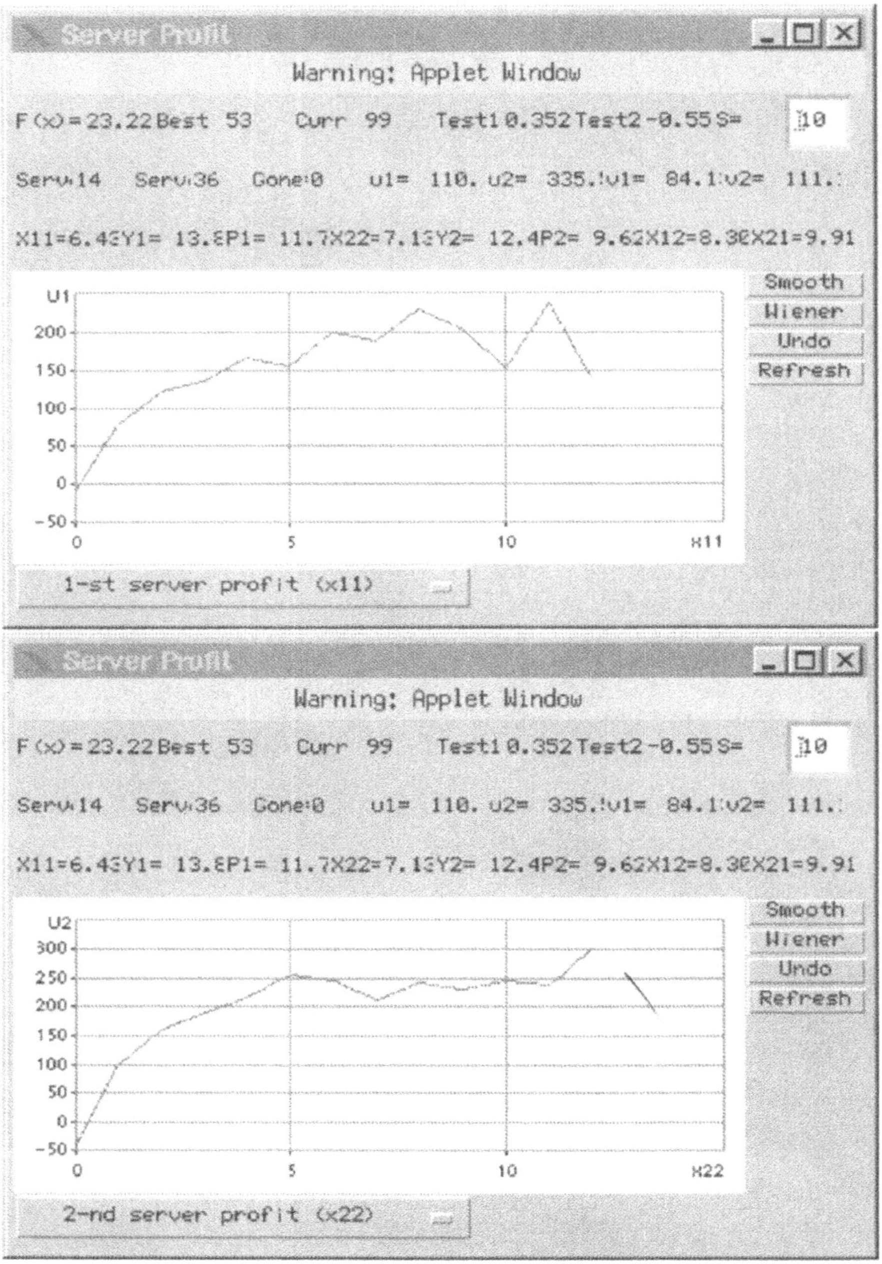

Figure 10.9. **Relation of the profit u_1 on the auxiliary resource x_{11} (top), Relation of the profit u_2 on the auxiliary resource x_{22} (bottom).**

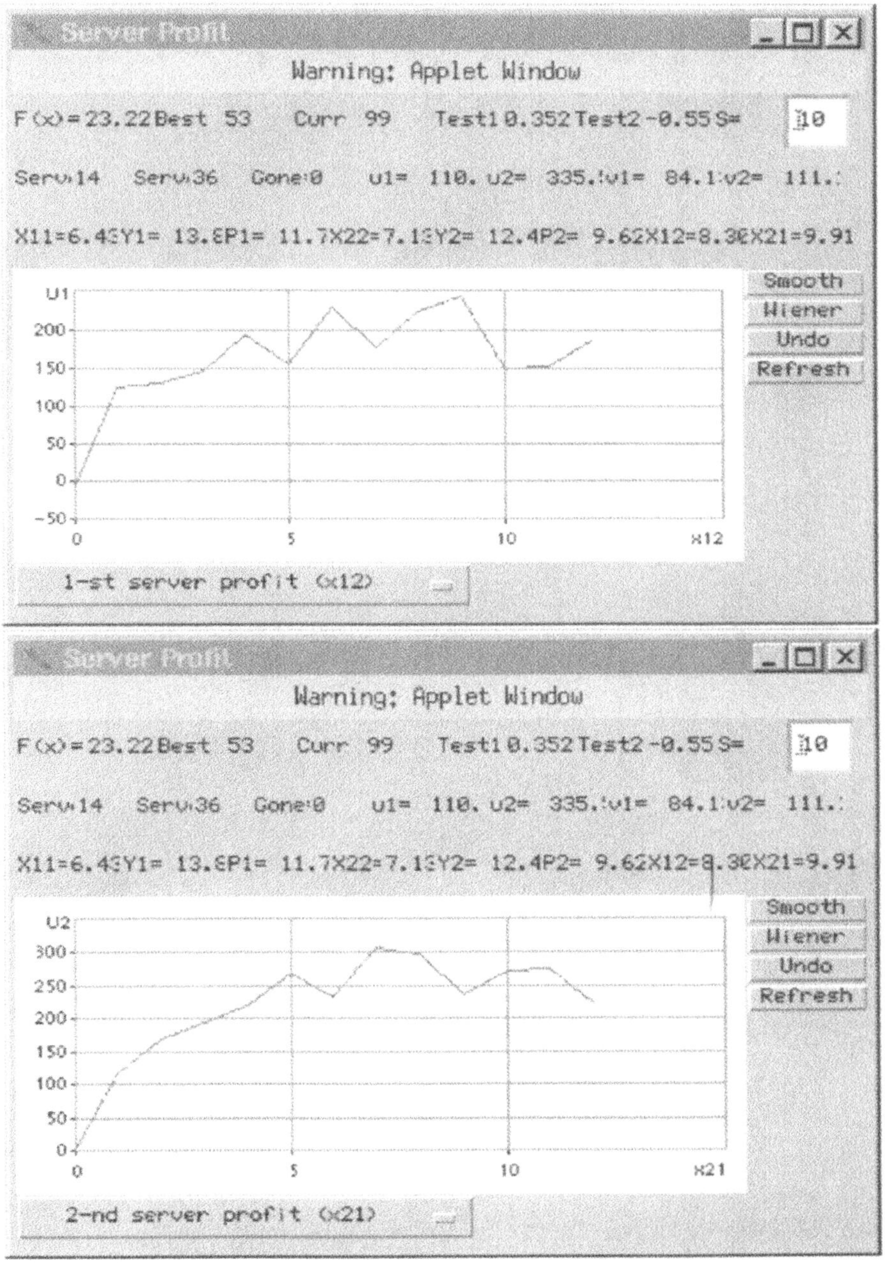

Figure 10.10. **Relation of the profit u_1 on the resources obtained from second server x_{12} (top), Relation of the profit u_2 on the resources obtained from first server x_{21} (bottom).**

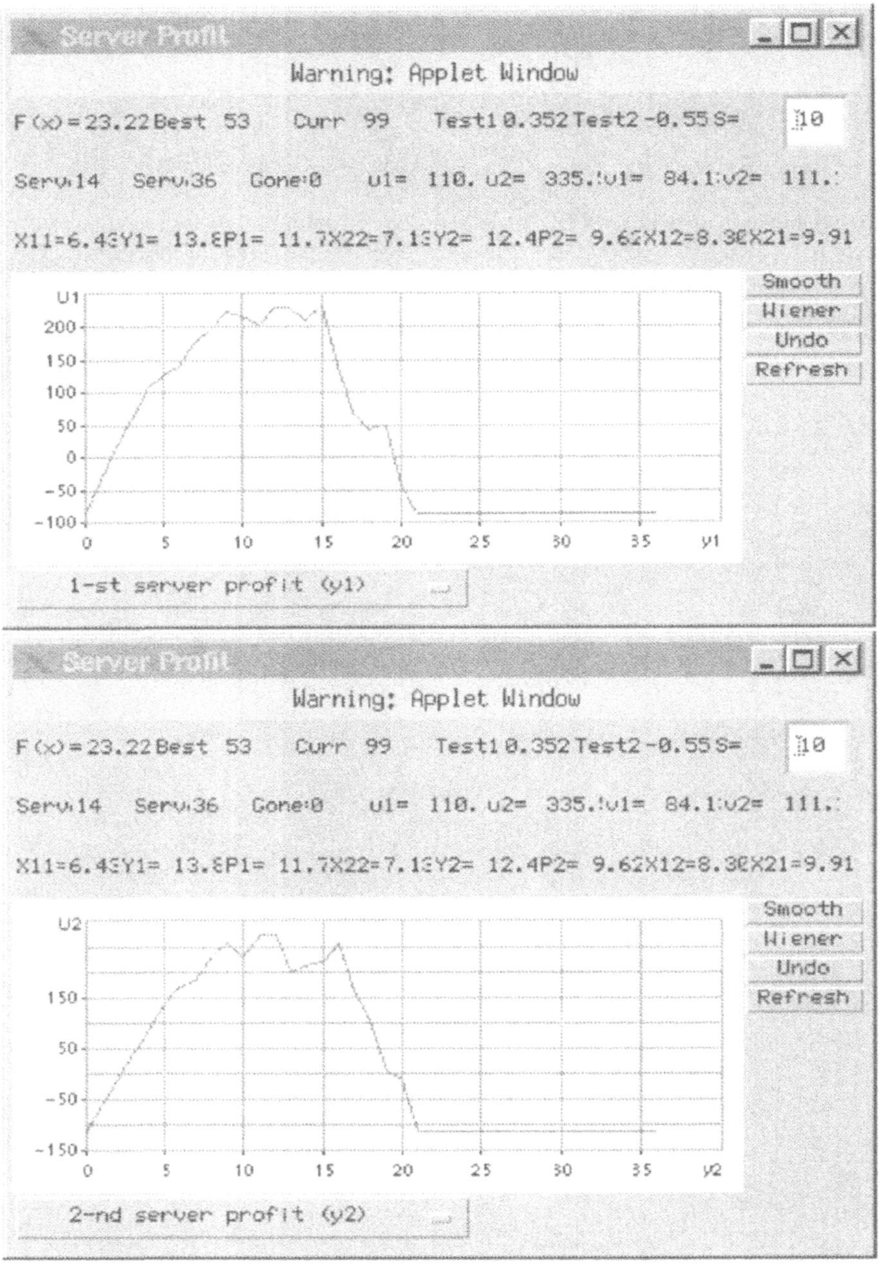

Figure 10.11. Relation of the profit u_1 on the service charge y_1 (top), Relation of the profit u_2 on the service charge y_2 (bottom).

Chapter 11

INSPECTION MODEL

The Nash and Walras models illustrate the competition in economics. Both models define such prices and other production parameters that satisfy the Nash equilibrium. However, the market competition represents only a part of economical and social activities. Another part is the inspection that one needs to provide tax collection, health and environment protection, e.t.c.. We consider a simple model that illustrates the competition between inspector and violator.

1. BIMATRIX GAME

Denote by $x = (x_1, ..., x_m),\ x_i \geq 0,\ \sum_i x_i = 1$ the inspection vector and by $y = (y_1, .., y_m),\ y_j \geq 0,\ \sum_j y_j = 1$ the violation vector. Here x_i denotes the inspection probability of the area i. y_j means the violation probability in the area j. Denote by $u(i, j)$ the inspection utility function when the object i is inspected and the object j is violated. Denote by $v(i, j)$ the violation utility function when the object i is inspected and the object j is violated. Functions $U(x, y)$ and $V(x, y)$ denote expected values of the inspection and violation utility functions using inspection and violation vectors x, y. These vectors define probabilities of inspection and violation. For example,

$$u(i, j) = \begin{cases} p_i g_i q_i, & \text{if } i = j \\ 0, & \text{otherwise,} \end{cases} \tag{11.1}$$

and

$$v(i,j) \quad = \quad \begin{cases} -q_j p_i g_j + (1-p_i) q_j g_j, & \text{if } i = j \\ q_j g_j, & \text{otherwise.} \end{cases} \tag{11.2}$$

Here p_i is the probability of detecting the violation, if it happens in the area i. q_i is the probability of completing the violation[1], if violation occurs in the area i. g_i is the utility of the completed violation in the area i.

Expression (11.1) means that if the violation is completed and detected[2] then the inspector premium is equal to the utility of violation[3]. Expression (11.2) shows that if the violation is completed and detected, the violator utility is negative. The utility is positive, if it is completed and not detected.

The average utility functions at fixed inspection and violation vectors x and y

$$U(x,y) = \sum_{i,j} x_i u(i,j) y_i, \tag{11.3}$$

and

$$V(x,y) = \sum_{i,j} x_i v(i,j) y_i. \tag{11.4}$$

Here $U(x,y) \neq V(x,y)$. Therefore, the inspection model is a bimatrix game (Forgo et al., 1999).

2. SEARCH FOR EQUILIBRIUM

Denote the contract-vector as $x^0 = (x_i^0, y_i^0,\ i = 1, ..., m)$. Then the fraud-vector $x^1 = (x_i^1, y_i^1,\ i = 1, ..., m)$, is obtained by maximizing the expected utilities $U(x,y)$ and $V(x,y)$ separately. It is assumed that the "partner" honors the contract $(x_i^0, y_i^0,\ i = 1, ..., m)$

$$x^1 = arg\max_x U(x, y^0),\ U = \max_x U(x, y^0), \tag{11.5}$$

$$y^1 = arg\max_y V(x^0, y),\ V = \max_y V(x^0, y). \tag{11.6}$$

[1] For example, killing prey.

[2] Prey is killed and a poacher is caught.

[3] The price of the killed prey.

These expressions define two linear programming problems that may be solved for any fixed contract vectors $x^0 = (x_i^0, y_i^0, \ i = 1, ..., m)$. However, using the standard simplex algorithm, one obtains the optimal base solution of linear programming problem, as usual. This solution represents the "best" vertex on some simplex. It follows from expressions (11.3)(11.4)(11.5) and (11.6) that the best vertex is defined by conditions

$$x_k^1 = \begin{cases} 1, & \text{if } \sum_j u(i,j)y_j < \sum_j u(k,j)y_j, k \neq i, \ i = 1, ..., m \\ 0, & \text{otherwise,} \end{cases}$$

and

$$y_l^1 = \begin{cases} 1, & \text{if } \sum_i v(i,j)x_i < \sum_i v(i,l)x_i, l \neq j, \ j = 1, ..., m \\ 0, & \text{otherwise,} \end{cases}$$

These conditions define a set of pure strategies (x^1, y^1). If the equilibrium point is in pure strategies, it can be obtained by the Direct Search Algorithm, too (see Section 2.1).

It is well known (Forgo et al., 1999) that, in randomly generated large bimatrix games, the proportion of games having k pure strategy equilibriums is defined by the Poisson distribution

$$e^{-1}/(k!) \tag{11.7}$$

That means that roughly two-thirds of randomly generated bimatrix games have at least one equilibrium point in pure strategies. This result holds for general randomly generated bimatrix games. In the inspector game, there is no equilibrium in pure strategies, as usual.

2.1 DIRECT SEARCH ALGORITHM (DSA)

A simple way to obtain equilibrium in pure strategies is the Direct Search Algorithm (DSA). The set IJ of equilibrium points in pure strategies is the intersection of two sets I and J defined by the following conditions

$$IJ = I \cap J, \tag{11.8}$$

$$I = \cup_j I(j), \ \ J = \cup_i J(i) \tag{11.9}$$

$$I(j), \subseteq arg\ max_i\ u(i,j) \tag{11.10}$$

$$J(i), \subseteq arg\ max_j\ v(i,j). \tag{11.11}$$

Here $I(j)$ is a set of maximal elements at each column of the matrix $u(i,j)$. $J(i)$ is a set of maximal elements at each row of the matrix

$v(i,j)$. If equilibrium exists and the intersection IJ is empty, one needs mixed strategies to obtain the equilibrium (Owen, 1968).

2.2 NECESSARY AND SUFFICIENT CONDITIONS

For a pair (x^0, y^0) to be an equilibrium point of the bimatrix game (A, B), it is necessary and sufficient (Forgo et al., 1999) that there exist real numbers α^0, β^0 such that $(x^0, y^0, \alpha^0, \beta^0)$ satisfies the system

$$\begin{aligned} xAx - \alpha &= 0, \\ xBy - \beta &= 0, \\ Ay - \alpha I &\leq 0, \\ xB - \beta I &\leq 0, \\ x \geq 0,\ y \geq 0,\ Ix &= 1, Iy = 1. \end{aligned} \tag{11.12}$$

Here I denotes the unit vector. The condition (11.12) represents the bilinear problem because it involves products of different variables. The solution of bilinear problems is difficult. Therefore, we consider two apprximate algorithms to search for mixed strategies of the equilibrium.

2.3 IRRELEVANT FRAUD ALGORITHM (IFA)

Often one obtains a mixed equilibrium strategy, if it exists, using a continuous set of all the solutions not just the base ones of the linear programming problems (11.5) and (11.6). Here the base solutions represent the pure strategies. The other solutions represent mixed strategies. The idea is to define such mixed strategies that makes the partner fraud irrelevant[4]. The following expressions define these strategies

$$\sum_{j=1}^{m} u(i,j) y_j^0 = U,\ i = 1, ..., m, \tag{11.13}$$

$$\sum_{i=1}^{m} v(i,j) x_i^0 = V,\ j = 1, ..., m, \tag{11.14}$$

[4]The fraud is irrelevant, if average winnings of the partner does not depend on his pure strategy $i = 1, ..., m$, and vice versa

$$\sum_{j=1}^{m} y_j^0 = 1, \tag{11.15}$$

$$\sum_{i=1}^{m} x_i^0 = 1, \tag{11.16}$$

$$x_i^0 \geq 0,\ y_j^0 \geq 0,\ i,j = 1, ..., m. \tag{11.17}$$

The expressions (11.13)- (11.17) define values of y_j^0 providing the multiple "continuous" maximum of $U(x, y^0)$. This means that any vector $x^1, \sum_i x_i^1 = 1,\ x_i \geq 0$ maximizes the average utility at given $y = y^0,\ y^0 = (y_1^0, ..., y_m^0)$,

$$x^1 = arg \max_x U(x, y^0),\ U = \max_x U(x, y^0), \tag{11.18}$$

and

$$y^1 = arg \max_y V(x^0, y),\ V = \max_y V(x^0, y). \tag{11.19}$$

Any probability distribution x^1 satisfies the maximum condition,. Therefore, one obtains the equilibrium just by setting

$$x^1 = x^0, \tag{11.20}$$

and

$$y^1 = y^0. \tag{11.21}$$

Expressions (11.13), (11.15, (11.17) include inequalities. Therefore, the linear programming (LP) is an appropriate method of solution. In LP all the variables are supposed to be nonnegative. Thus, variables U and V should be expressed as differences of two nonnegative variables $U = u_1 - u_2$ and $V = v_1 - v_2$. Then from expressions (11.13) (11.15 (11.17) one obtains this LP problem

$$\max_{x,y,u,v} (u_1 - u_2 + v_1 - v_2), \tag{11.22}$$

$$\sum_{j=1}^{m} u(i,j) y_j = u_1 - u_2,\ i = 1, ..., m, \tag{11.23}$$

$$\sum_{i=1}^{m} v(i,j) x_i = v_1 - v_2,\ j = 1, ..., m, \tag{11.24}$$

$$\sum_{j=1}^{m} y_j^0 = 1, \tag{11.25}$$

$$\sum_{i=1}^{m} x_i^0 = 1, \quad (11.26)$$

$$x_i \geq 0,\ y_j \geq 0,\ i,j = 1,...,m, u_1 \geq 0,\ u_2 \geq 0, v_1 \geq 0, v_2 \geq 0. \quad (11.27)$$

Here $x = (x_1, ..., x_m)$, $y = (y_1, ..., y_m)$, $u = (u_1, u_2)$, $v = (v_1, v_2)$. The solution of this linear programming problem exists not always. Then one searches for the equilibrium in pure strategies using the Direct Search algorithm (11.8). That complements the linear programming algorithm (11.25), (11.26),(11.27) and helps to define the equilibrium. An alternative is the strategy elimination algorithm.

2.4 STRATEGY ELIMINATION ALGORITHM (SEA)

If no equilibrium is detected by both these algorithms, then following procedure is used:

1. obtain a solution (x^*, y^*) of linear equations (11.13),(11.14),(11.15), and (11.16), while ignoring inequalities (11.17),
2. set to zero all negative variables: $x_i = 0,\ if\ x_i^* \leq 0;\ y_j = 0,\ if\ y_j^* \leq 0$,
3. reduce the system of linear equations by eliminating columns j and raws i with zero variables $j : y_j = 0$ and $i : x_i = 0$,
4. if a solution of the reduced system is found, stop, and record the zero componentsas pure strategies, record the positive components $x_i > 0,\ y_j > 0$ as mixed equilibrium strategies,
5. if the reduced system is not empty, go to step 1.

This is the Strategy Elimination Algorithm (SEA). Sometimes (SEA) obtains solutions of bilinear problems (11.12). The algorithm defines both the pure and the mixed strategies. Therefore, SEA may replace both the Direct Search Algorithm (11.8) and the linear programming one (11.25), (11.26),(11.27).

3. MATRIX GAME

In some special cases, the Inspector problem is reduced to the Matrix Game. It is well known (Owen, 1968) that equilibrium is defined as two linear programming problems, if $v_{ij} = -u_{ij}$. One reduces utilities (11.1) and (11.2) to the zero-sum case by subtracting the penalty terms $(1-p_i)q_jg_j$ and q_ig_i from the inspectors utilities. In such a case

$$u(i,j) = \begin{cases} p_ig_iq_i - (1-p_i)q_jg_j, & \text{if } i = j \\ -q_ig_i, & \text{otherwise,} \end{cases} \tag{11.28}$$

and

$$v(i,j) = \begin{cases} -q_jp_ig_j + (1-p_i)q_jg_j, & \text{if } i = j \\ q_jg_j, & \text{otherwise.} \end{cases} \tag{11.29}$$

It follows form expressions (11.28) and (11.29) that $v_{ij} = -u_{Ij}$. Here one obtains the equilibrium by solving two linear programming problems (Owen, 1968)

$$\max_x U, \tag{11.30}$$

$$\sum_{i=1}^m x_iu_{ij} \geq U, \; j = 1, ..., m,$$

$$\sum_{i=1}^m x_i = 1, \; x_i \geq 0,$$

$$U = u_1 - u_2, \; u_1 \geq 0, \; u_2 \geq 0,$$

and

$$\min_y V, \tag{11.31}$$

$$-\sum_{j=1}^m y_ju_{ij} \leq V, \; i = 1, ..., m,$$

$$\sum_{i=1}^m y_i = 1, \; y_i \geq 0,$$

$$V = v_1 - v_2, \; v_1 \geq 0, \; v_2 \geq 0.$$

The important feature the zero-sum games is the mini-max condition

$$\min_y V = \max_x U. \tag{11.32}$$

Chapter 12

"DUEL" PROBLEM, DIFFERENTIAL GAME MODEL

1. INTRODUCTION

In all three competition models, the Nash, the Walras, and the Inspector, the steady-state conditions were considered, for simplicity. In real life competition processes, dynamics are essential. Dynamical competition models may be represented as differential games. They are difficult for numerical and visual analysis, as usual (Friedman, 1971; Galperin and Zheng, 1991; Hajek, 1975; Isaacs, 1965; Krasovski, 1970; von Neumann and Morgenstern, 1953; Petrosyan, 1985; Petrosyan, 1993; Rodin, 1987). Therefore, one tries to make the dynamic competition models as simple as possible.

Consider, as a simple illustration, two objects in space trying to destroy each other. Assume, as a first approximation, that the control parameters are: initial points z_0, w_0, "climbing" rates a, b and firing times t_1, t_2. These parameters are set before the start for both objects. Denote the control parameters by vectors $x = (x_i,\ i = 1, 2, 3)$ and $y = (y_i,\ i = 1, 2, 3)$, correspondingly. Denote by (a_i^x, a_i^y) the lower bounds of these parameters. Denote by (b_i^x, b_i^y) the upper bounds. In the illustrative example $x_1 = z_0, x_2 = a, x_3 = t_1$, and $y_1 = w_0, y_2 = b, y_3 = t_2$, where $-1 \leq z_0 \leq 1, -1 \leq w_0 \leq 1, -1 \leq a \leq 1, -1 \leq b \leq 1$. Movements of the objects are described in two-dimensional "hight-time" space by the first-order differential equations

$$dz(t)/dt = az(t), \tag{12.1}$$

$$dw(\tau)/d\tau = bw(\tau),\ \tau = 2 - t. \tag{12.2}$$

The solution of these equations defines trajectories of the objects

$$z(t) = z_0 e^{at}, \tag{12.3}$$
$$w(\tau) = w_0 e^{bt}. \tag{12.4}$$

As usual, models of moving objects use second-order differential equations, at least

$$d^2 z(t)/dt^2 + g z(t)/dt = a z(t), \tag{12.5}$$
$$d^2 w(t)/dt^2 + h dw(\tau)/d\tau = b w(\tau), \ \tau = 2 - t. \tag{12.6}$$

We apply first-order linear differential equations (12.1) and (12.2) as a first approximation to test algorithms of optimization and modeling. Denote by $d(t)$ a distance between objects at the moment t

$$d(t) = ||(w(t), \tau) - (z(t), t)||. \tag{12.7}$$

Denote by $p(t)$ the hitting probability

$$p(t) = 1 - (d(t)/D)^{\alpha}. \tag{12.8}$$

Here $D \geq \max_t d(t)$ and $\alpha > 0$. The expected utility function of the first object

$$U(x, y) = \begin{cases} u_1 p(x_3) - u_2(1 - p(x_3))p(y_3), & \text{if } x_3 < y_3, \\ -u_2 p(y_3) + u_1(1 - p(y_3))p(x_3), & \text{if } x_3 > y_3, \\ u_1 p(x_3) - u_2 p(y_3), & \text{if } x_3 = y_3, \end{cases} \tag{12.9}$$

The expected utility function of the second object

$$V(x, y) = \begin{cases} v_1 p(y_3) - v_2(1 - p(y_3))p(x_3), & \text{if } x_3 > y_3, \\ -v_2 p(x_3) + v_1(1 - p(x_3))p(y_3), & \text{if } x_3 < y_3, \\ v_1 p(y_3) - v_2 p(x_3), & \text{if } x_3 = y_3. \end{cases} \tag{12.10}$$

Expressions (12.9) and (12.10) of expected utilities follows from expression (12.8) of hitting probabilities. It is assumed that utilities to hit the hostile object are plus u_1 and v_1. Utilities to be hit are minus u_2 and v_2. Here u_1, u_2 denote utilities of the first object and v_1, v_2 define utilities of the second one.

2. CONVEX VERSION

Expected utilities (12.9) and (12.10) are not convex functions of variables $p(x_3)$ and $p(y_3)$. The convex version may be obtained assuming that the object destroys the enemy, if it is not hit itself. Suppose that $u_1 = u_2 = v_1 = v_2 = 1$. Then it follows from expressions (12.9) and (12.10) that the expected utility function of the first object

$$U(x,y) = \begin{cases} p(x_3) - (1 - p(x_3)), & \text{if } x_3 < y_3, \\ -p(y_3) + (1 - p(y_3)), & \text{if } x_3 > y_3, \\ p(x_3) - p(y_3), & \text{if } x_3 = y_3, \end{cases} \tag{12.11}$$

The expected utility function of the second object

$$V(x,y) = \begin{cases} p(y_3) - (1 - p(y_3)), & \text{if } x_3 > y_3, \\ -p(x_3) + (1 - p(x_3)), & \text{if } x_3 < y_3, \\ p(y_3) - p(x_3), & \text{if } x_3 = y_3. \end{cases} \tag{12.12}$$

Expected utilities (12.9) and (12.10) are convex functions of probabilities $p(x_3)$ and $p(y_3)$ at the equilibrium point

$$p(x_3) = 0.5,$$
$$p(y_3) = 0.5. \tag{12.13}$$

The convexity of expected utilities (12.9) and (12.10) as functions of probabilities $p(x_3)$ and $p(y_3)$ does not provide their convexity as functions of firing times x_3 and y_3 and other parameters. The convexity is a part of sufficient equilibrium conditions. Therefore, the equilibrium may not exist.

Expected utilities reach the maximal values at the ends of intervals of parameters z_0, a, w_0, b, as usual. For the first object there are four pure strategies corresponding to various arrangements of these ends
$(z_0 = z_{min}, a = a_{min}), (z_0 = z_{min}, a = a_{max}), (z_0 = z_{max}, a = a_{min}),$
$(z_0 = z_{max}, a = a_{max})$.
The pure strategies for the second object:
$(w_0 = w_{min}, b = b_{min}), (w_0 = w_{min}, b = b_{max}), (w_0 = w_{max}, b = b_{min}),$
$(w_0 = w_{max}, b = b_{max})$.
We may search for the pure equilibrium of bimatrix games (12.11)(12.12) corresponding to each fixed pair $(x_3.y_3)$ of firing times using simple discrete algorithm (11.8). However, the probability to obtain the pure equilibrium is high only for large bimatrix games (Forgo et al., 1999). Therefore, we consider mixed strategies, first.

3. MIXED STRATEGIES

If expected utilities reach the maximal values at the ends of intervals of parameters z_0, a, w_0, b then mixed strategies could be helpful. The mixed

strategies mean that each object chooses at random a combination of lower and upper bounds. Define those bounds as zero-ones for simplicity. There are four events:
$(z_0 = 0, a = 0), (z_0 = 0, a = 1), (z_0 = 1, a = 0), (z_0 = 1, a = 1)$.
Probabilities of these events are defined by first four components
$x(1), ..., x(4),\ x(i) \geq 0,\ \sum_{i=1}^{4} x(i) = 1$ of a vector
$x = (x(i),\ i = 1, ..., 5)^1$.
The fifth component $x(5) = t_1$ defines the firing time. A vector
$y = (y(j),\ j = 1, ..., 5)$ has a similar meaning for the second object.

Applying mixed strategies $x = (x_i,\ i = 1, ..., 5)$ and
$y = (y_j,\ j = 1, ..., 5)$, the utility functions of the first and the second objects are

$$U_a(x, y) = \sum_{r \in R} \sum_{s \in S} x(r) u(r, s) y(s), \tag{12.14}$$

$$V_a(x, y) = \sum_{r \in R} \sum_{s \in S} x(r) u(r, s) y(s). \tag{12.15}$$

Here $R = \{1, 2, 3, 4\}$, $S = \{1, 2, 3, 4\}$, where $1, 2, 3, 4$ denotes the four events $(z_0 = 0, a = 0), (z_0 = 0, a = 1), (z_0 = 1, a = 0), (z_0 = 1, a = 1)$. The matrix

$$u(r, s) = u_{t_1, t_2}(r, s) \tag{12.16}$$

is the winning of the first object, if the pure strategy denoted by the quadruple (t_1, t_2, r, s) is used. The matrix

$$v(r, s) = v_{t_1, t_2}(r, s) \tag{12.17}$$

is the winning of the second object, if the pure strategy (t_1, t_2, r, s) is used. Expanding expressions (12.14)

$$\begin{aligned} U_a(x, y) = {} & x(1)u(1,1)y(1) + x(2)u(2,1)y(1) + \\ & x(3)u(3,1)y(1) + x(4)u(4,1)y(1) + \\ & x(1)u(1,2)y(2) + x(2)u(2,2)y(2) + \\ & x(3)u(3,2)y(2) + x(4)u(4,2)y(2) + \\ & x(1)u(1,3)y(3) + x(2)u(2,3)y(3) + \end{aligned}$$

[1]The variables x, y obtain different meanings, if one uses mixed strategies. That reduces the number of symbols. To avoid confusion, the utilities $U_a(x, y), V_a(x, y)$ are supplied with the subscript a. If the subscript a is absent then x, y defines the parameters z_0, w_0, a, b, t_1, t_2. Otherwise, x, y denotes mixed strategies.

$$x(3)u(3,3)y(3) + x(4)u(4,3)y(3) +$$
$$x(1)u(1,4)y(4) + x(2)u(2,4)y(4) +$$
$$x(3)u(3,4)y(4) + x(4)u(4,4)y(4). \qquad (12.18)$$

Here winnings $u(r,s)$ are defined by expression (12.16). Expression (12.15) illustrates how the function $U_a(x,y)$ is defined. The function $V_a(x,y)$ is defined similarly.

One defines bimatrix games (12.14) and (12.15) for each pair of firing times (t_1, t_2). Therefore, for fixed pair (t_1, t_2), the same methods of search for the equilibrium points may be used as in the inspector model (11.8) and (11.13)- (11.17). Here winnings $u(r,s)$ and $v(r,s)$ depend on firing times t_1, t_2. This is the important difference from the inspector game. One updates bimatrix games while changing firing times.

Therefore, there are only two parameters t_1 and t_2 to be found by the fraud minimization (see section 4.1). One can do that by applying methods, such as the *Bayes*, to search for the pair (t_1, t_2) that minimizes the fraud.

4. SEARCH FOR EQUILIBRIUM

4.1 BIMATRIX GAME ALGORITHM

Consider the bimatrix game (12.14) and (12.15) defined by fixing a pair (t_1, t_2) of firing times. We search for the equilibrium of this game using conditions (11.8) and (11.13)- (11.17). Here the matrix is small thus the proportion of pure strategy equilibrium points will not be considerable (Forgo et al., 1999). Therefore we start by looking for such a mixed strategy that makes all the partner strategies equally profitable for him. That makes the partner fraud irrelevant. At fixed pair (t_1, t_2) a mixed strategy $x(r)$ of the first object is "fraud irrelevant" if

$$\sum_{r \in R} x(r)v(r,s) = V, \ s \in S, \qquad (12.19)$$
$$\sum_{r \in R} x(r) = 1, \ x(r) \geq 0.$$

There are four linear inequalities, five linear equations and five variables: V, $x(r)$, $r \in R$, because the set R has four elements.

A mixed strategy $y(s)$ of the second object is fraud irrelevant if

$$\sum_{s\ inS} u(r,s)y(s) = U, \ r \in R, \tag{12.20}$$

$$\sum_{s \in S} y(s) = 1.y(s) \geq 0.$$

These expressions include inequalities what is usual in the linear programming (LP). Applying LP, the variables U and V should be expesed as differences of two non-negative variables $U = u_1 - u_2$ and $V = v_1 - v_2$. Then from the expressions (12.19), (12.20) one obtains the LP problem:

$$\max_{x,y,u,v} (u_1 - u_2 + v_1 - v_2) \tag{12.21}$$

$$\sum_{j=1}^{m} u(i,j)y_j = u_1 - u_2, \ i = 1,...,m, \tag{12.22}$$

$$\sum_{i=1}^{m} v(i,j)x_i = v_1 - v_2, \ j = 1,...,m, \tag{12.23}$$

$$\sum_{j=1}^{m} y_j = 1, \tag{12.24}$$

$$\sum_{i=1}^{m} x_i = 1 \tag{12.25}$$

$$x_i \geq 0, \ y_j \geq 0, \ i,j = 1,...,m, u_1 \geq 0, \ u_2 \geq 0, v_1 \geq 0, v_2 \geq 0.. \tag{12.26}$$

Here $x = (x_1,...,x_m), \ y = (y_1,...,y_m), \ u = (u_1,u_2), \ v = (v_1,v_2)$.

Similar linear fraud irrelevant equations were considered in the inspection problem (see expressions (11.23) and (11.24)). Here the fraud irrelevant equations (12.22) and (12.23) depend on two unknown parameters t_1 and t_2. Therefore, linear problems (12.22) and (12.23) are just a part of general non-linear problem. We use longer notations, such as $U = U(t_1,t_2)$ and $V = V(t_1,t_2)$, to show that U, V depend on t_1, t_2.

If there is no solution of fraud irrelevant equations, we search for the pure equilibrium strategies using condition (11.8). The Strategy Elimination Algorithm (SEA) is a good alternative to obtain equilibrium. A general but difficult way to find the equilibrium is to solve the bilinear problem (11.12) defining the necessary and sufficient equilibrium conditions.

Obtaining equilibrium in mixed strategies $x(r), y(s), \ r \in R, \ s \in S$ or in pure strategies r, s, one reduces the six-dimensional optimization problem to the two-dimensional one.

$$\min_{z \in B} \| z - T_t(z) \|^2 . \tag{12.27}$$

Here $z = (t_1^0, t_2^0)$, where (t_1^0, t_2^0) are the "contract" firing times. The operator T_t is defined by the fraud condition

$$t_1^1 = arg \max_{t_1} U(t_1, t_2^0),$$
$$t_2^1 = arg \max_{t_2} V(t_1^0, t_2). \tag{12.28}$$

Here $U(t_1, t_2^0)$ is the expected winning of the first object. It is the solution of the corresponding bimatrix game at fixed firing times t_1 and t_2^0. $V(t_1^0, t_2)$ is the expected winning of the second object defined as the solution of the bimatrix game at given firing times t_1^0 and t_2. For example, winnings $U(t_1, t_2^0)$ and $V(t_1^0, t_2)$ can be defined by LP problems (12.22)(12.23), or by the Direct Search Algorithm (DSA) (see Section 2.1), or by Strategy Elimination Algorithm (SEA) (see Section 2.4).

The equilibrium is achieved, if the minimum (12.27) is zero. The minimization of (12.27) is the task for global optimization algorithms, including the Bayesian ones.

4.2 MATRIX GAME ALGORITHM

The bimatrix game algorithm can be applied to both the symmetric and the non-symmetric cases. In the symmetric case the utilities satisfy the zero-sum game condition $U = -V$. Here one obtains the equilibrium by solving the following linear programming problems

$$\max_x U \tag{12.29}$$

$$\sum_{i=1}^{m} x_i u_{ij} \geq U, \; j = 1, ..., m,$$

$$\sum_{i=1}^{m} x_i = 1, \; x_i \geq 0$$

$$U = u_1 - u_2, \; u_1 \geq 0, \; u_2 \geq 0,$$

and

$$\min_y V, \tag{12.30}$$

$$-\sum_{j=1}^{m} y_j u_{ij} \leq V, \ i = 1, ..., m,$$

$$\sum_{i=1}^{m} y_i = 1, \ y_i \geq 0$$

$$V = v_1 - v_2, \ v_1 \geq 0, \ v_2 \geq 0.$$

Here mini-max condition (Owen, 1968)

$$\min_y V = \max_x U \tag{12.31}$$

holds for all firing times t_1, t_2. Therefore, a search for the equilibrium is reduced to the condition

$$\min_{t_1, t_2} (U^2 + V^2). \tag{12.32}$$

The zero-sum equilibrium is reached when the hitting probabilities are equal to 0.5 for both players and their winnings are zero (see the one-dimensional example in the next section).

5. ONE-DIMENSIONAL EXAMPLE

Applying expressions (12.1)-(12.10) in one-dimensional space the trajectories are

$$z(t) = t, \tag{12.33}$$

$$w(\tau) = 2 - \tau, \tag{12.34}$$

and the hitting probability is

$$p(t) = 1 - d(t)/D. \tag{12.35}$$

Here $D \geq \max_t d(t)$. The control parameters are the firing times $x = t_1$ and $y = t_2$. Assuming that $D = 2$ and applying expression (12.11), the expected utility function of the first object is

$$U(x,y) = \begin{cases} x - (1 - x), & \text{if } x < y, \\ -y + (1 - y), & \text{if } x > y, \\ x - y, & \text{if } x = y. \end{cases} \tag{12.36}$$

The expected utility function of the second object follows from expression (12.12)

$$V(x,y) = \begin{cases} y - (1 - y), & \text{if } x > y, \\ -x + (1 - x), & \text{if } x < y, \ . \\ y - x, & \text{if } y = x, \end{cases} \tag{12.37}$$

The equilibrium is reached at the firing moment $x = y = 0.5$. The one-dimensional example helps to understand and to test the results of the two-dimensional case.

6. ECONOMIC DUEL. NASH MODEL

By the "Economic Duel" we mean a dynamic case of the Nash model. In Chapter 9, the static case was described.

There are m servers providing the same service

$$u_i = u_i(x_1, y_1, ..., x_m, y_m) = a_i y_i - x_i, \ i = 1, ..., m,$$

where u_i is the profit, y_i is the service price, a_i is the rate of customers, x_i is the running cost, and i is the server index

Suppose that arriving customers estimate the average waiting time as the relation of the number of waiting customers n_i to the server capacity x_i that is assumed to be equal to the running cost

$$\gamma_i = n_i / x_i$$

A customer goes to the server i, if

$$c_i \leq c_j, \ j = 1, ..., m, \ j \neq i, \ c_i \leq c_0.$$

A customer goes away, if

$$\min_i c_i > c_0,$$

where c_0 is the critical cost. The rate a of incoming consumers is fixed

$$a = \sum_{i=0}^{m} a_i,$$

To illustrate general ideas of the dynamic Nash model, we simplify the static one, first.

6.1 SIMPLIFIED NASH MODEL. STATIC CASE

We express the customer estimate of the expected waiting time as

$$\gamma_i = a_i / x_i, \ i = 1, 2. \tag{12.38}$$

Assume the customer balance condition.

$$a_i/x_i + y_i = q. \tag{12.39}$$

Here q is the customer balance factor. The balance condition follows from (9.4),(9.5), (9.6), if the number of customers is large. The condition is based on the observation that arriving customers prefer servers with lesser sum $y_i + \gamma_i$. From expression (9.7)

$$q \leq c. \tag{12.40}$$

From (12.39)

$$a_i = (q - y_i)x_i, \tag{12.41}$$

and

$$u_i = (q - y_i)x_i y_i - x_i. \tag{12.42}$$

Consider a monopolistic situation when only the ith server is there. In this case

$$a_i = a, \ if \ \ q \leq c \tag{12.43}$$

and

$$y_i + a/x_i \leq c. \tag{12.44}$$

Othervise, customers abandon services $a_i = 0$. The maximal monopolistic profit

$$\max_{x_i, y_i}(ay_i - x_i), \tag{12.45}$$

$$y_i + a/x_i \leq c. \tag{12.46}$$

Replacing inequality (12.46) by equality, we eliminate (12.46) and maximize the expression

$$max_{y_i}(ay_i - a/(c - y_i)). \tag{12.47}$$

Now one defines the optimal service price

$$y_i^* = c - 1, \tag{12.48}$$

the optimal service capacity,

$$x_i^* = a, \tag{12.49}$$

and the maximal monopolistic profit

$$u_i^* = a(c - 2). \tag{12.50}$$

6.2 DYNAMIC NASH MODEL

In the previous section, considering the static model, it was tacitly assumed that a server can operate steadily while making loss instead of profit. Now we consider a case of insolvency, too. Suppose that a server gets broken, if the accumulated losses exceed some credit threshold. Here the bankrupt server is eliminated and the remaining one assumes a monopolistic position.

Denote by $U_i(t)$ a profit accumulated by the ith server at a time t. Assume that

$$U_i(t) = \int_{t_0}^{t} u_i(t)dt, \tag{12.51}$$

where $u_i(t)$ defines a profit at a moment t

$$u_i(t) = a_i(t)y_i(t) - x_i(t). \tag{12.52}$$

Here $a_i(t)$ is a rate of customers of the ith server at a moment t. This rate is formally defined as a limit of the fraction

$$a_i(t) = \lim_{\delta \to 0} \frac{A_i(t+\delta) - A_i(t)}{\delta}, \tag{12.53}$$

where $A_i(t)$ is a total number of customers arrived during an interval $(0,t)$, $t \le T$. The zero denotes the starting time of the system. T is the end of the operation period, the "horizon". In real life $A_i(t)$ is defined at fixed moments, for example, minutes, hours, days e.t.c.. We consider it as a continuous parameter because we consider a continuous model.

Denote the bankruptcy threshold of an ith server by U_i^*. Denote a bankruptcy moment (if it happens) of the ith server by t_i^*. Then

$$a_i(t) \quad = \quad \begin{cases} a, & \text{if } t \ge t_j^*,\ t \le t_i^*, i,j = 1,2,\ i \ne j, \\ 0, & \text{if } t \le t_j^*,\ t \ge t_i^*\ i,j = 1,2,\ i \ne j. \end{cases} \tag{12.54}$$

Otherwise, customer rates $a_i(t)$ are defined by the customer balance condition (12.41)[2] .

From (12.50) the profit

$$u_i(t) \quad = \quad \begin{cases} u_i^*, & \text{if } t \ge t_j^*,\ t \le t_i^*, i,j = 1,2,\ i \ne j, \\ 0, & \text{if } t \le t_j^*,\ t \ge t_i^*\ i,j = 1,2,\ i \ne j. \end{cases} \tag{12.55}$$

---

[2]This means that the customer balance follows the changes of parameters $y_i(t)$ and $x_i(t)$ almost without delay.

Otherwise, the profit $u_i(t)$ is defined by (12.52).

Assume that both servers are planning their dynamic competition in advance. Each server i defines trajectories of the service price $y_i(t)$ and the service capacity $x_i(t)$. These trajectories show how the service prices and capacities depend on the time $t \in (0, T)$. Objectives are to maximize profits

$$U_i = U_i(T), \ i = 1, 2. \tag{12.56}$$

Here $U_i(T)$ defines the accumulated profit (12.51) of ith server at the end of period $t = T$.

We illustrate the idea by a simple example. Assume that servers control initial values $y_i(0)$, $x_i(0)$ and rates of change b_{y_i}, b_{x_i} of parameters $y_i(t)$ $x_i(t)$. Then the trajectories are defined by differential equations

$$dy_i(t)/dt = b_{y_i} y_i(t), \tag{12.57}$$

and

$$dx_i(t)/dt = b_{x_i} x_i(t). \tag{12.58}$$

The corresponding solutions are

$$y_i(t) = y^i(0) \exp(b_{y_i} y_i(t)), \tag{12.59}$$

and

$$x_i(t) = x^i(0) \exp(b_{x_i} x_i(t)). \tag{12.60}$$

SEARCH FOR NASH EQUILIBRIUM

The search is similar to that in the Nash model (see Section 9). We fix the the initial values, the "Contract-Vector"
$z^0 = (x_i(0)^0, b^0_{x_i}, y_i(0)^0, b^0_{y_i}, \ i = 1, ..., m)$. The transformed values, the "Fraud-Vector"
$z^1 = (x_i(0)^1, b^1_{x_i}, y_i(0)^1, b^1_{y_i}, \ i = 1, 2)$, is obtained by maximizing the profits of each server i. The maximization is performed assuming that all partners $j \neq i$ honors the contract
$(x_i(0)^0, b^0_{x_j}, y_i(0)^0, b^0_{y_j}, \ j = 1, 2 \ j \neq i)$

$$(x_i(0)^1, b^1_{x_i}, y_i(0), b^1_{y_i}) =$$

$$arg \max_{x_i(0), b_{x_i}, y_i(0), b_{y_i}} U_i(x_i(0), b_{x_i}, y_i(0), b_{y_i}, x_j(0)^0, b^0_{x_j}, y_i(0)^0, b^0_{y_j},$$

$$j = 1, 2 \ j \neq i), \ i = 1, 2. \tag{12.61}$$

Here the symbol U_i denotes the accumulated profit (12.56). Formally, condition (12.61) transforms the vector
$z^n = ((x_i(0)^n, b^n_{x_i}, y_i(0)^n, b^n_{y_i},\ i = 1,2) \in B \subset R^8,\ n = 0,1,2,...$ into the vector z^{n+1}. To make expressions shorter denote this transformation by T

$$z^{n+1} = T(z^n),\ n = 0,1,2,... \tag{12.62}$$

One may obtain the equilibrium at the fixed point z^n, where

$$z^n = T(z^n). \tag{12.63}$$

The fixed point z^n exists, if the feasible set B is convex and all the profit functions (9.1) are convex (Michael, 1976). We obtain the equilibrium directly by iterations (12.62) , if the transformation T is contracting. If not, then we minimize the square deviation

$$\min_{z \in B} \parallel z - T(z) \parallel^2 . \tag{12.64}$$

The equilibrium is achieved, if the minimum (12.64) is zero.

One makes the model more flexible by introducing second derivatives of parameters $y_i(t)$ and $x_i(t)$. The second derivatives represent accelerations. Assume that servers control the starting values $y_i(0),\ x_i(0)$, the rates $b_{y_i},\ b_{x_i}$ and accelerations $g_{y_i},\ g_{x_i}$ of parameters $y_i(t)\ x_i(t)$. Then trajectories are defined by differential equations

$$d^2y_i(t)/dt^2 = g_{y_i}dy_i(t)/dt + b_{y_i}y_i(t), \tag{12.65}$$

and

$$d^2x_i(t)/dt^2 = g_{x_i}dx_i(t)/dt + b_{x_i}x_i(t). \tag{12.66}$$

Here the vector z^n has twelve components.
$z^n = (x_i(0)^n, b^n_{x_i}, g^n_{x_i}, y_i(0)^n, b^n_{y_i}, g^n_{y_i}\ i = 1,2) \in B \subset R^{12},\ n = 0,1,2,...$
That means high dimension of the fraud minimization problem (12.64). In addition, one needs to test the convexity of accumulated profits U_i defined by expressions (12.56) as functions of parameters
$x_i(0), b_{x_i}, y_i(0), b_{y_i},\ i = 1,2$. If they are convex then one obtains equilibrium by condition (12.64). Otherwise, some specific algorithms should be designed. An example is the inspector problem.

7. SOFTWARE EXAMPLE

An interactive duel model is implemented using Java JDK1.2 (Matulevicius, 1999). To run the Java applet go to web-sites (see section 4.),

start $GMJ2$ and select the task *DuelStarwars* (see Figure 8.6).
The task window, in Figure 12.1, shows the domain of firing times $0 \leq t_1 \leq 1$, $0 \leq t_2 \leq 1$.

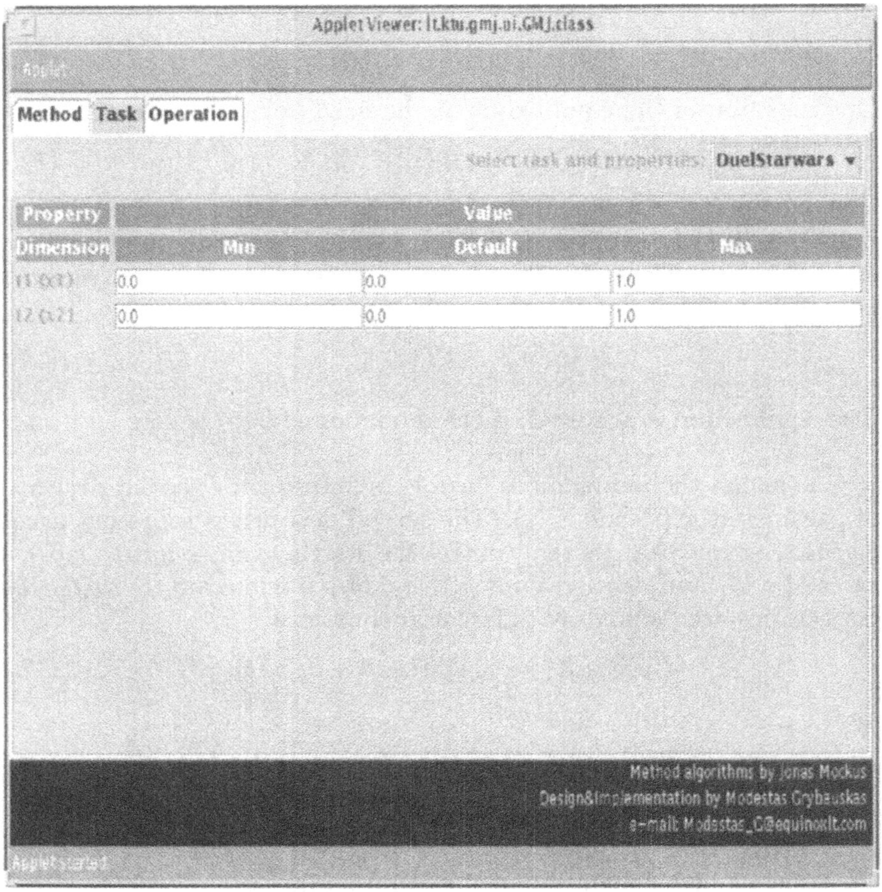

Figure 12.1. The duel task window.

One selects the optimization method from the *Methods* menu (see Figure 8.7). The method *Bayes* is selected (see Figure 12.2). The total number of iterations is set to forty. The number of initial iterations is set to five.

The optimization results are in Figure 12.3.
To start the analysis of results one touches the *Analysis* button (see Figure 8.9).

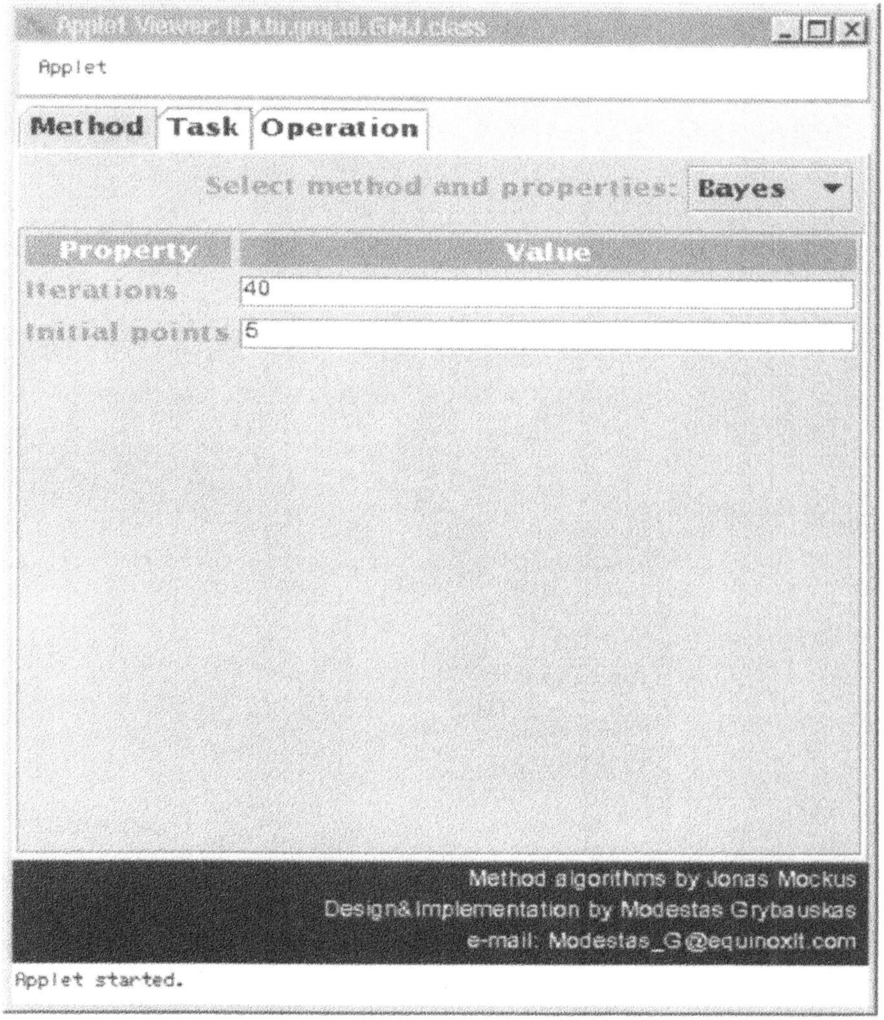

Figure 12.2. **The Bayes method selection.**

The *Convergence* mode shows how the deviation from the equilibrium depends on the iteration number (see Figure 12.4).

7.1 OPERATING DUEL

The duel starts by selecting *DuelStarwars* mode in the *Analysis* menu (see Figure 8.8).

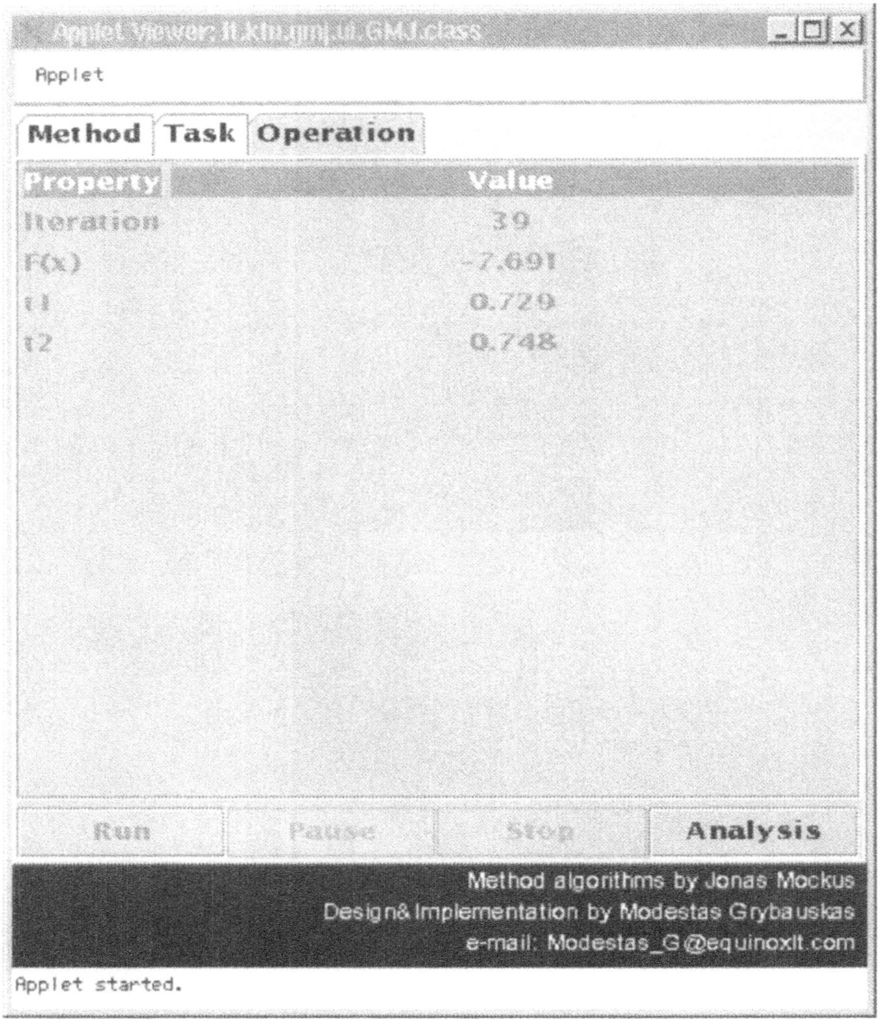

Figure 12.3. The results.

To run duel press the *Start* button (see Figure 12.5).

One selects the operation mode in the operation window (see Figure 12.6). There are three operation modes:

- Person vs Person,
- Person vs Computer,
- Computer vs Computer.

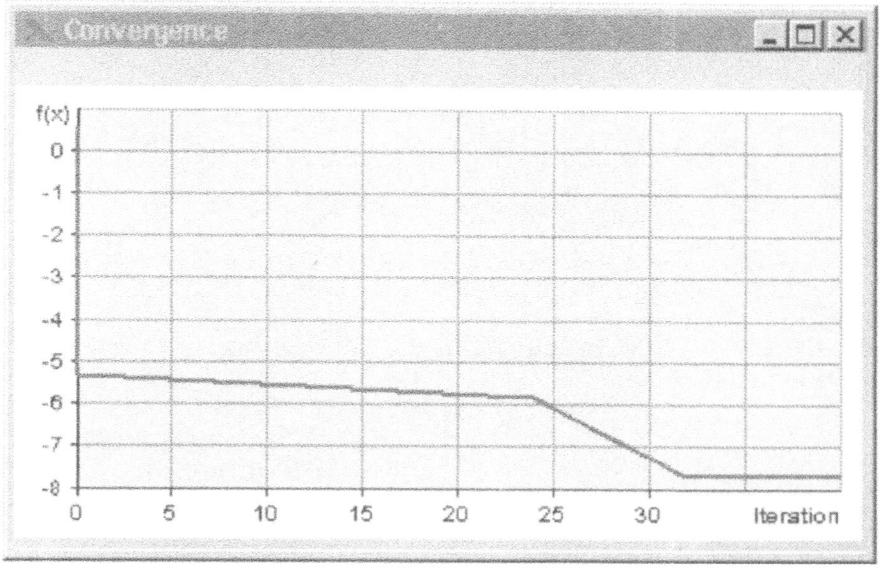

Figure 12.4. The convergence window.

Figure 12.6 shows the *Computer vs Computer* mode. The *Current situation* field shows parameters z_0, w_0, a, b generated at random using probabilities defined by equilibrium conditions. The firing times t_1, t_2 are decided by these conditions, directly.

In the *Person vs Computer* game the person fix parameters z_0, a, t_1 of the first object. A window to fix them is in the center of the top Figure 12.7.

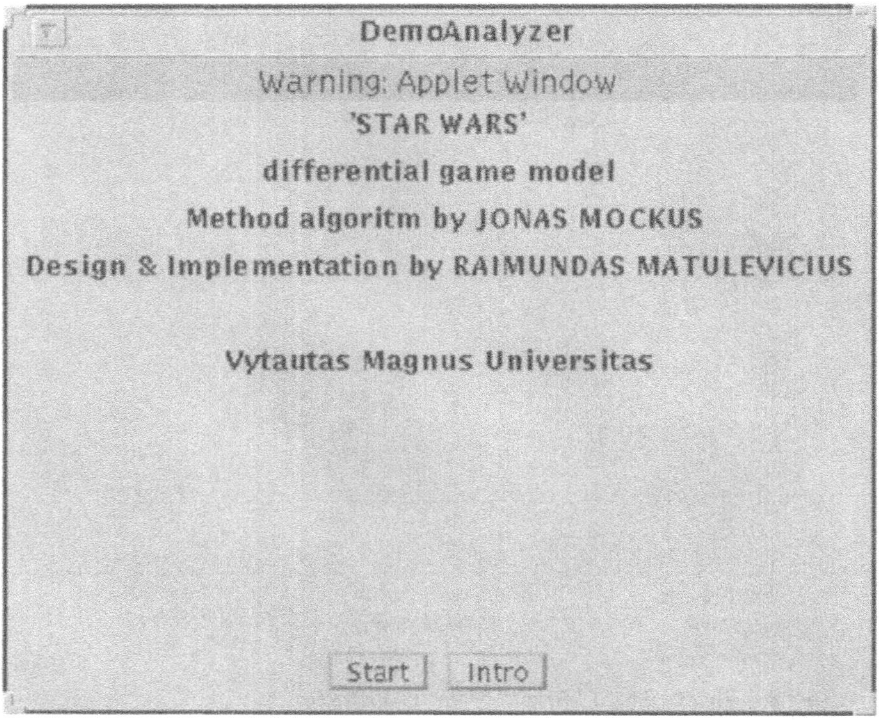

Figure 12.5. The starting window.

Probabilities of parameters w_0, b of the second object and its firing time t_2 are defined by equilibrium conditions (12.24),(12.23), and (12.27). Parameters w_0, b of the second object, and the actual hitting, are defined by probabilities. Therefore, the outcome is random.

Figure 12.8 shows the input of the *Person vs Person* game. In this game parameters of both objects are fixed by persons using the input window. The outcome of this game is shown in a form similar to the *Person vs Computer* game (see the bottom Figure 12.7).
One sets the speed of moving objects in the *Speed* window (see Figureduelout1).
Figures 12.10 show *help* windows.

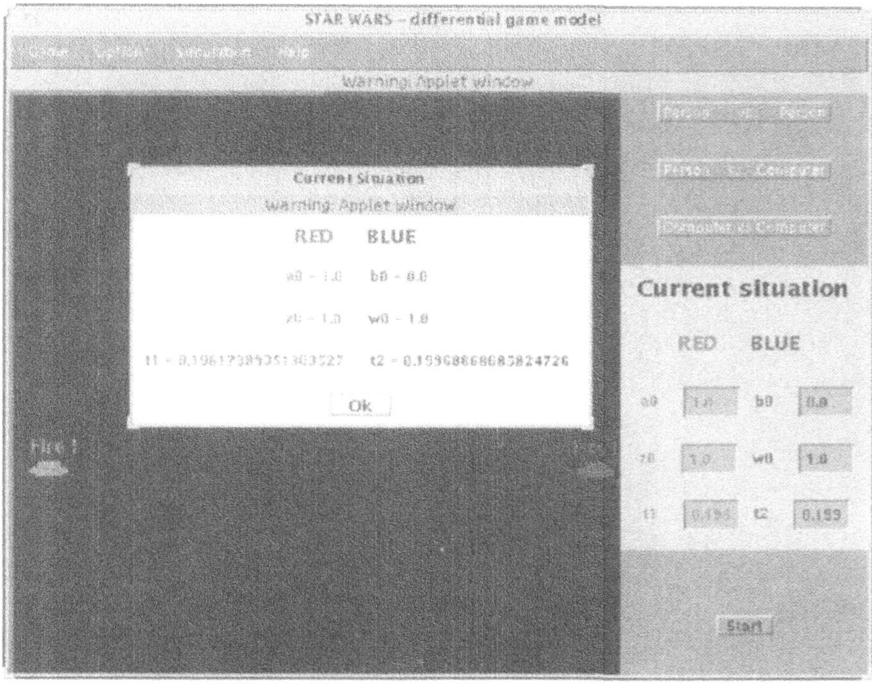

Figure 12.6. The operation window in the *Computer vs Computer* game.

7.2 FUTURE DEVELOPMENTS

In the example, one searches for equilibrium in mixed strategies using the Irrelevant Fraud Algorithm (IFA) which is described in section 2.3. This algorithm solves linear programming problems (12.22) and (12.23). If no solution is found, one searches for pure equilibrium strategies using the Direct Search Algorithm (DSA) (11.8). The Strategy Elimination Algorithm (SEA) (see section 2.3) helps, if both IFA and DSA fail.

A general but difficult way is to use the necessary and sufficient equilibrium conditions given as bilinear problem (11.12). This is a future task. The main future task is to implement flexible economic models. Another task is the implementation of second-order models. These tasks need additional theoretical investigations and software developments. Creation of entertaining graphical interfaces is important, too.

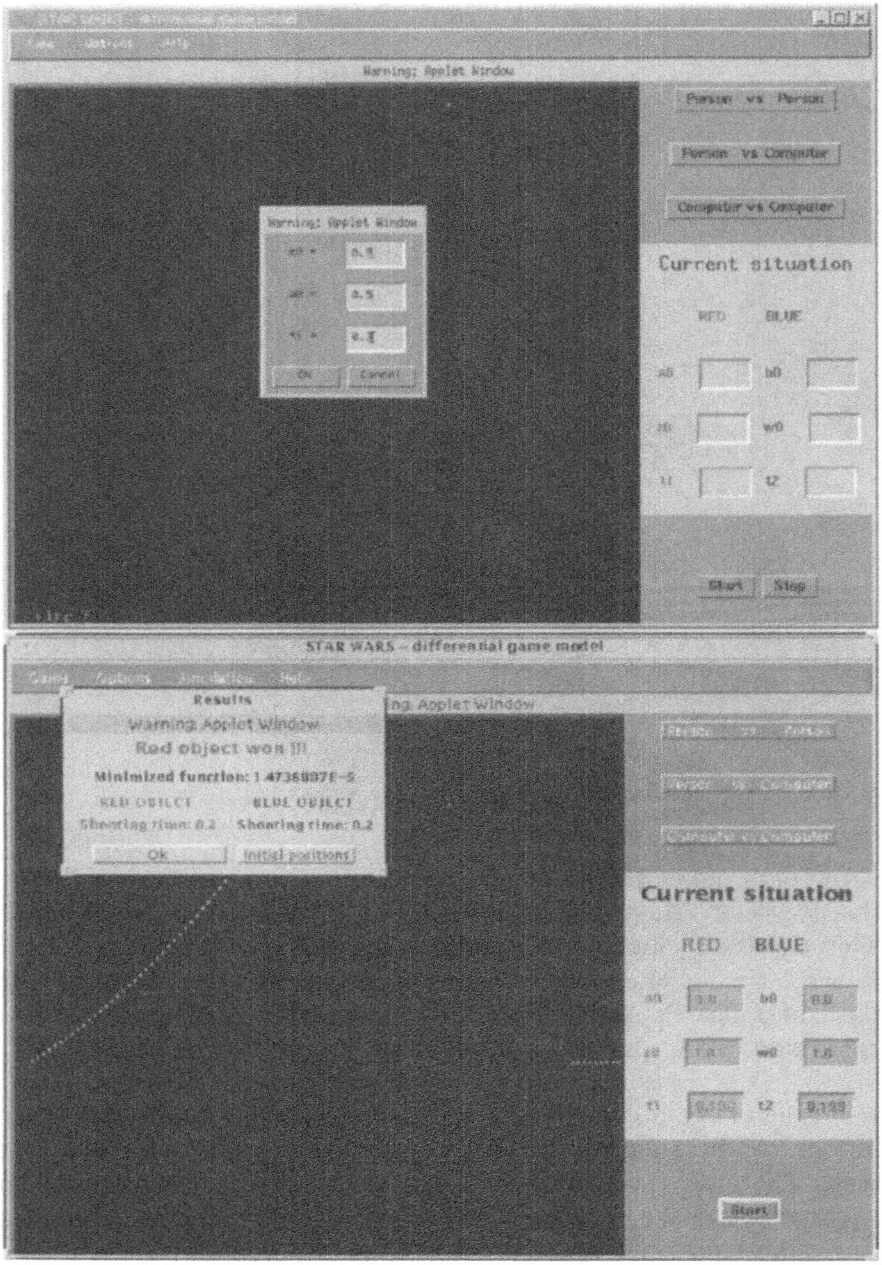

Figure 12.7. The input (top) and the outcome (bottom) of the *Person vs Computer* game.

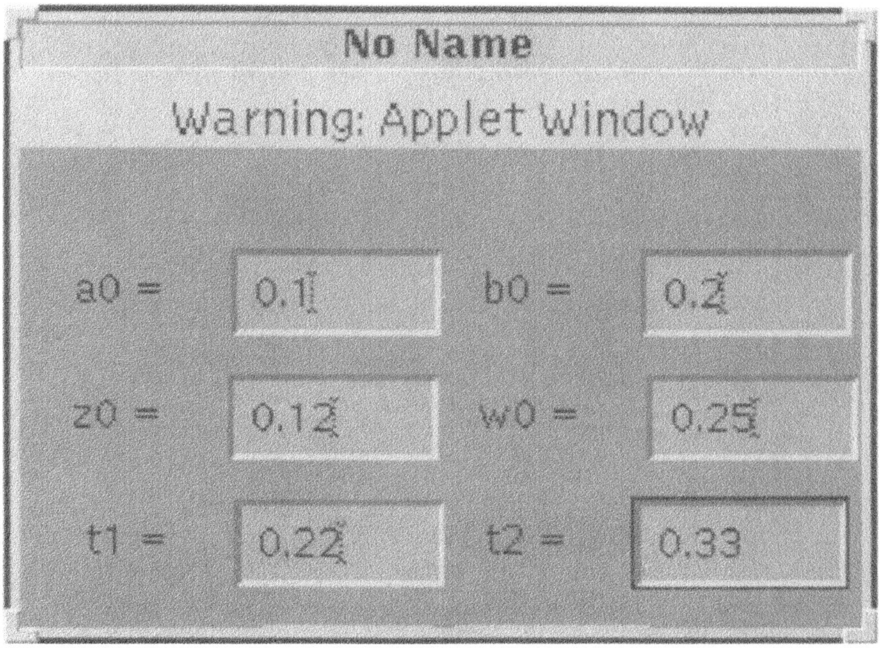

Figure 12.8. The input of the ***Person vs Person*** game.

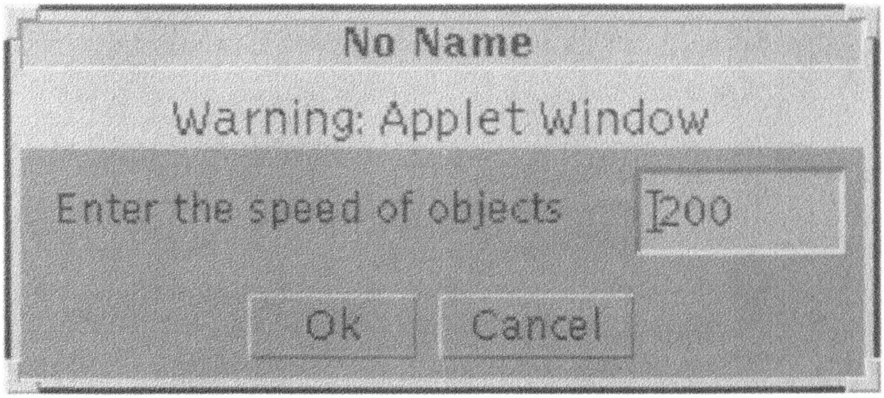

Figure 12.9. The speed control window.

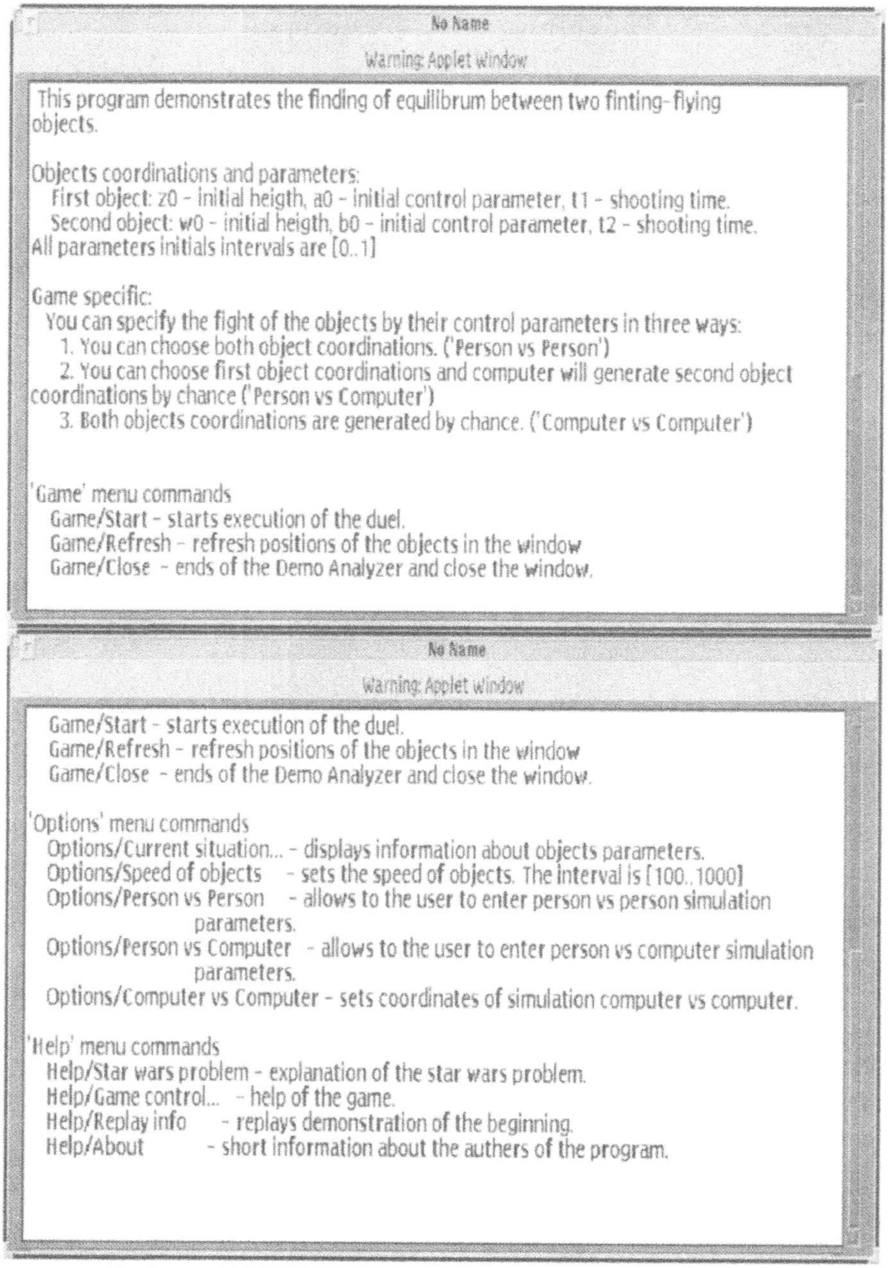

No Name

Warning: Applet Window

This program demonstrates the finding of equilibrum between two finting-flying objects.

Objects coordinations and parameters:
First object: z0 - initial heigth, a0 - initial control parameter, t1 - shooting time.
Second object: w0 - initial heigth, b0 - initial control parameter, t2 - shooting time.
All parameters initials intervals are [0..1]

Game specific:
You can specify the fight of the objects by their control parameters in three ways:
1. You can choose both object coordinations. ('Person vs Person')
2. You can choose first object coordinations and computer will generate second object coordinations by chance ('Person vs Computer')
3. Both objects coordinations are generated by chance. ('Computer vs Computer')

'Game' menu commands
Game/Start - starts execution of the duel.
Game/Refresh - refresh positions of the objects in the window
Game/Close - ends of the Demo Analyzer and close the window.

No Name

Warning: Applet Window

Game/Start - starts execution of the duel.
Game/Refresh - refresh positions of the objects in the window
Game/Close - ends of the Demo Analyzer and close the window.

'Options' menu commands
Options/Current situation... - displays information about objects parameters.
Options/Speed of objects - sets the speed of objects. The interval is [100..1000]
Options/Person vs Person - allows to the user to enter person vs person simulation parameters.
Options/Person vs Computer - allows to the user to enter person vs computer simulation parameters.
Options/Computer vs Computer - sets coordinates of simulation computer vs computer.

'Help' menu commands
Help/Star wars problem - explanation of the star wars problem.
Help/Game control... - help of the game.
Help/Replay info - replays demonstration of the beginning.
Help/About - short information about the authers of the program.

Figure 12.10. The ***help*** pages.

Chapter 13

"PORTFOLIO" PROBLEM, OPTIMAL INVESTMENT OF RESOURCES

1. INTRODUCTION

The previous examples illustrated competition and inspection processes in economical, social and ecological problems. Here the optimal investment of available resources is considered. Investment problems depend on the nature of resources to be invested. An important part of any investment problem is a proper definition of utility functions that determine the profit-to-risk relation. Here we consider an illustrative example how to invest some fixed capital in Certificates of Deposit (CD) and Stocks.

The portfolio problem is to maximize the average utility of a wealth. That is obtained by optimal distribution of available capital between different objects with uncertain parameters (Mockus et al., 1997). Denote by x_i the part of the capital invested into an object i. The returned wealth is $y_i = c_i x_i$. Here $c_i = 1 + \alpha_i$ and $\alpha_i > 0$ is an interest rate. Denote by $p_i = 1 - q_i$ the reliability of investment. Here q_i is the insolvency probability. $u(y)$ is the utility the wealth y. Denote by $U(x)$ the expected utility function. $U(x)$ depends on the capital distribution $x = (x_i, ..., x_n)$, $\sum_i = 1$, $x_i \geq 0$. If y is continuous, the expected utility function

$$U(x) = \mathbf{E}u(y) = \int_0^\infty u(y)p(y)dy. \tag{13.1}$$

Here $p(y)$ is probability density of wealth y. If the wealth is discrete $y = y^k,\ k = 1, ..., M$, the expected utility function

$$U(x) = \sum_{k=1}^{M} u(y^k)p(y^k). \tag{13.2}$$

Here M is the number of discrete values of wealth y^k. $p_x(y^k)$ is the probability that the wealth y^k will be returned, if the capital distribution is x. We search for such capital distribution x which provides the greatest expected utility of the returned wealth:

$$\max_x U(x), \tag{13.3}$$

$$\sum_{i=1}^{n} x_i = 1, \tag{13.4}$$

$$x_i \geq 0.$$

2. EXPECTED UTILITY

One may define probabilities $p(y^j)$ of discrete values of wealth y^j, $j = 1, 2, ...$ by exact expressions. For example,

$$\begin{aligned}
p(y^1) &= p_1 \prod_{i \neq 1} q_i, \\
p(y^2) &= p_2 \prod_{i \neq 2} q_i, \\
\dots\dots\dots &\quad \dots\dots\dots\dots\dots\dots \\
p(y^n) &= p_n \prod_{i \neq n} q_i, \\
p(y^{n+1}) &= p_1 p_2 \prod_{i \neq 1, i \neq 2} q_i, \\
p(y^{n+2}) &= p_1 p_3 \prod_{i \neq 1, i \neq 3} q_i \\
\dots\dots\dots &\quad \dots\dots\dots\dots\dots\dots
\end{aligned} \tag{13.5}$$

Then from expression (13.5)

$$U(x) = \sum_{k=1}^{M} u(y^k)p(y^k). \tag{13.6}$$

Here M is the number of different values of wealth y.

One determines $U(x)$ approximately by the Monte Carlo approach:

$$U_K(x) = 1/K \sum_{k=1}^{K} u(y^k). \tag{13.7}$$

Here

$$y^k = \sum_{i=1}^{n} y_i^k, \tag{13.8}$$

where

$$y_i^k = \begin{cases} c_i x_i, & \text{if } \eta_i^k \in [0, p_i] \\ 0, & \text{otherwise.} \end{cases} \tag{13.9}$$

Here K is the number of Monte Carlo samples. η_i^k is a random number uniformly distributed on the unit interval. In this case

$$U(x) = \lim_{K \to \infty} U_K(x). \tag{13.10}$$

3. OPTIMAL PORTFOLIO, SPECIAL CASES

The optimal portfolio depends on the utility function $u(y)$. Consider, for example, the optimal portfolio for three different utility functions.

The first utility function is linear

$$u(y) = cy. \tag{13.11}$$

This function is for "rich" persons. Rich persons want to maximize the average wealth. They are not emotional about accidental losses or gains. In the linear case (13.11), the optimal portfolio is to invest all the capital in an object with the highest product $p_i c_i$.

The second utility function is for "prudent" persons which averse risk

$$u(y) = \begin{cases} 0, & \text{if } 0 \le y < a \\ 1, & \text{if } a \le y \le c, \end{cases} . \tag{13.12}$$

Here a is a risk threshold. $c = \max_i c_i x_i$ denotes the maximal return of invested capital (see expression (13.4)). If $a = 1/m \ min_i \ c_i x_i$ then, in the risk-averse case (13.12), an optimal decision is $x_i^* = 1/m, \ i = 1, ..., m$. Here one divides the capital equally between all the objects[1].

[1]The optimal decision $x = 1/m$ is not unique, any decision satisfying the inequality $c_i x_i \ge a, \ i = 1, ..., m$ minimizes the expected utility function $U(y)$.

The third utility function is for "risky" persons. Risky persons are ready to risk for the great win c.

$$u(y) = \begin{cases} 0, & \text{if } 0 \leq y < c, \\ 1, & \text{if } y = c. \end{cases} \tag{13.13}$$

Here one invests all the capital in the object with highest wealth return. Therefore, $x_i = 1$, if $c_i = \max_j c_j = c$.

These examples are abstract. An average person behaves "risky," if only a small part of his resources is involved. The same person behaves prudently, if all his wealth is at stake. There is a point r between areas of risky and prudent behavior. At this point an average person behaves like the "rich" one. Here is an example

$$\begin{aligned} u(y) &< y, \ if \ 0 \leq y < r, \\ u(y) &= y, \ if \ y = r, \\ u(y) &> y, \ if \ r < y \leq c. \end{aligned} \tag{13.14}$$

Here r is a boundary point between risky and prudent areas.

4. UTILITY FUNCTIONS

Utility functions $u(y)$ are different for different persons and organizations. An individual utility function is defined by a lottery $L(A, B, p) = \{pA + (1-p)B\}$. Here p is the probability to win the best event A.
$(1-p)$ is the probability to get the worst one B. Denote by C the "ticket price" of this lottery. There are two possible decisions:

- keep the ticket money C,
- bay a ticket and risk losing this money while hoping to win a greater wealth A with probability p.

Denote by $p(C)$ a "hesitation" probability, when one cannot decide which decision to prefer. One defines the "hesitation" probability $p(C)$ by this condition

$$L(A, B, C, p(C)) = [C \approx \{p(C)A + (1 - p(C))B\}]. \tag{13.15}$$

Here the symbol $\approx$ denotes the "hesitation." If utilities $u(A) = 1$ and $u(B) = 0$, the utility of the "ticket" C is equal to the hesitation probability $u(C) = p(C)$ (Fishburn, 1964).

Suppose, for example, that event C is to keep all the investment capital, $y = 1$, in a safe; no risk, no profit. Assume that the event A means doubling the capital, $y = 2$. The event B means losing all the capital, $y = 0$.

Denote by $p(1)$ the hesitation probability. Then $u(1) = u(0) + p(1)(u(2) - u(0))$. If $u(0) = 0$ and $u(2) = 1$ then the utility of the capital $u(1) = p(1)$. Here one obtained capital utilities at three points: $y = 0$, $y = 1$, and $y = 2$.

To define a reasonable approximation of the utility function $u(y)$, we need at least two additional points. For example, points $y = 0.5$ and $y = 1.5$. One defines the corresponding utilities by the hesitation probabilities $p(0.5)$ and $p(1.5)$. These are obtained by two hesitation lotteries

$$L(1.0, 0.0, 0.5, p(0.5)) = \\ [(y = 0.5) \approx \{p(0.5)(y = 1) + (1 - p(0.5))(y = 0)\}] \tag{13.16}$$

and

$$L(2.0, 1.0, 1.5, p(1.5)) = \\ [(y = 1.0) \approx \{p(1.5)(y = 2.0) + (1 - p(1.5))(y = 1)\}]. \tag{13.17}$$

Here one obtains utility values $u(0) = 0$, $u(0.5) = p(0.5)$, $u(1) = p(1)$, $u(1.5) = u(1) + p(1.5)(u(2) - u(1))$ $u(2) = 1$. The remaining capital utility values are defined by the linear interpolation

$$u(y) = u(y_i) + p(y_i)(u(y_{i+1}) - u(y_i)), \quad y_i \leq y < y_{i+1}, \\ i = 0, 1, ..., 4. \tag{13.18}$$

In consulting offices, the "psychological tests" defining capital utilities are not always convenient. Then one of the four "typical" utility functions can be selected. The typical utility functions represent the risky, the average, the rich and the prudent persons. The selection depends on observable personal traits.

The same capital utility function (13.14) could be used for all customers. Then one defines customer differences by different border points, namely:
$r_{prudent} < r_{average} < r_{rich} < r_{risky}$ (see Figure 13.2).

5. SOFTWARE EXAMPLE

5.1 RUNNING THE PROGRAM

The software (Petrikis and Juodgalvyte, 1997) is on the web-site (see Section 4.) and is run by Internet

1. open the line 'GMJ1' on the 'Software systems' page (see Figure 3.1),
2. start the applet,
3. open the 'Methods' menu, choose the method and its parameters (see Figure 13.1),
4. open the 'Tasks' menu, chose the task 'Portfolio',
5. select the character : prudent, rich, risky, average (see Figure 13.2),
6. select the method: Monte Carlo, Deterministic (see Figure 13.2),
7. fix the predicted stock rates by updating or keeping the default values determined using ARMA model (see Figure 13.2),
8. click at the 'Operation' label at the top of page,
9. click at the 'Run' button at the bottom of page,
10. read results of the optimal investment, where 'Iteration' means the number of iterations to reach the optimal investment, 'F(x)' shows the expected utility with the minus sign, the following lines show the relative investments[2] (see Figure 13.3),
11. open the 'convergence' window (see Figure 13.4) which show how the best utility depends on the iteration number,
12. open the 'projection' window (see Figure 13.5) showing how the utility depends on the first variable,
13. open the 'spectrum' window showing how a frequency of utilities depends on the iteration number[3].

The example of predicted stock rate of Vilnius Bankas is presented in the file 'onestep?VB.txt'.

[2]The actual investment is the product of the total investment and the relative investment, divided by the sum of all the relative investments.

[3]Different frequencies are is denoted by different colors.

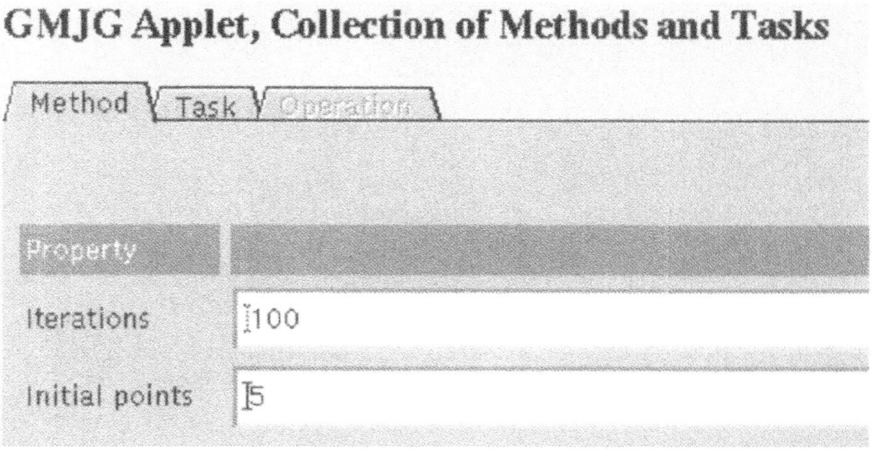

Figure 13.1. The control window.

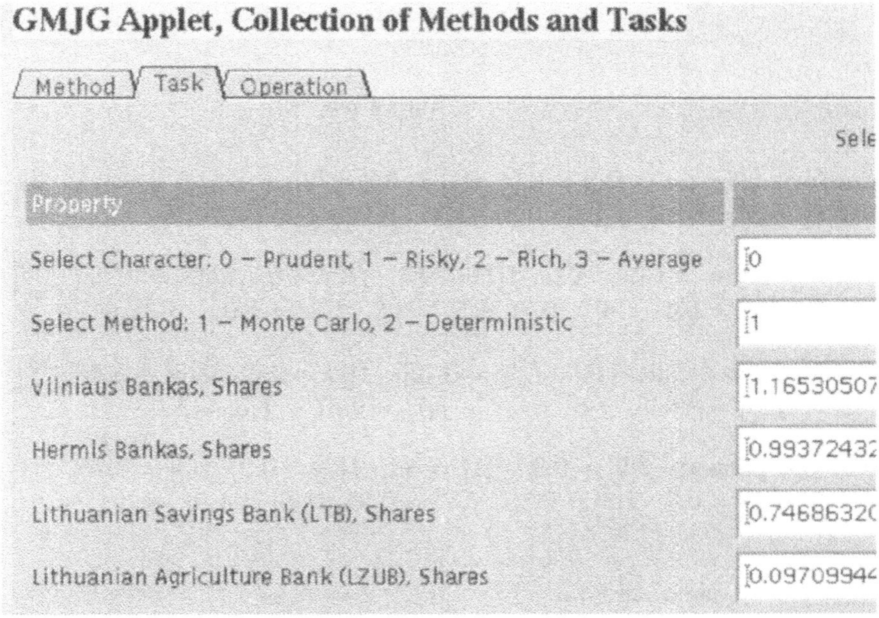

Figure 13.2. The input fields.

5.2 A SET OF UTILITY FUNCTIONS

Nine approximation points are used: $yk[i] = 0.25 * i,\ i = 0, ..., 8$. Four utility functions are available. They differ at the approximation points. They represent different persons.

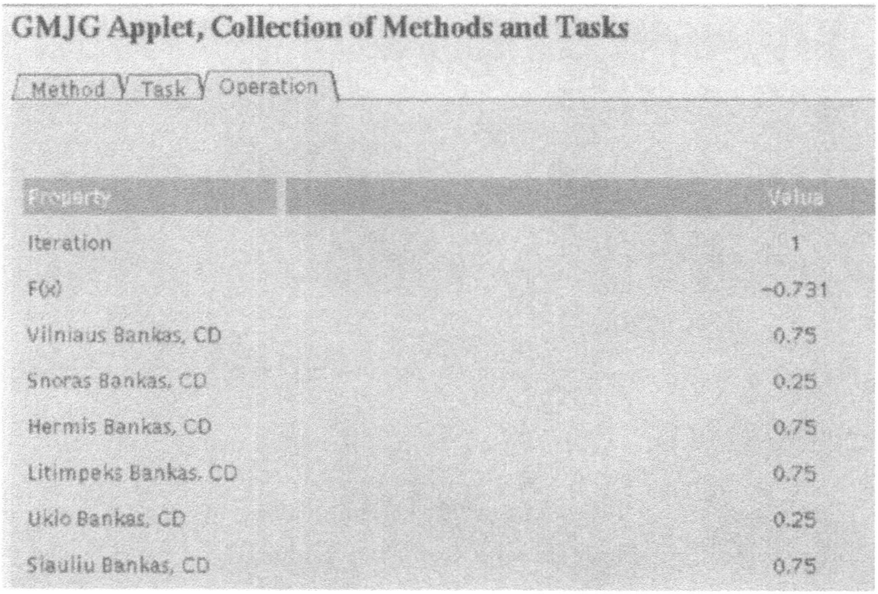

Figure 13.3. The output field.

prudent person : $f[0] = 0.0,\ f[1] = 0.3,\ f[2] = 0.5,\ f[3] = 0.7,\ f[4] = 0.8,\ f[5] = 0.85,\ f[6] = 0.9,\ f[7] = 0.95,\ f[8] = 1.0,$

risky person : $f[0] = 0.0,\ f[1] = 0.1,\ f[2] = 0.2,\ f[3] = 0.25,\ f[4] = 0.3,\ f[5] = 0.4,\ f[6] = 0.7,\ f[7] = 0.9,\ f[8] = 1.0,$

rich person : $f[0] = 0.0,\ f[1] = 0.125,\ f[2] = 0.25,\ f[3] = 0.325,\ f[4] = 0.5,\ f[5] = 0.625,\ f[6] = 0.75,\ f[7] = 0.875,\ f[8] = 1.0,$

average person $f[0] = 0.0,\ f[1] = 0.1,\ f[2] = 0.2,\ f[3] = 0.25,\ f[4] = 0.5,\ f[5] = 0.7,\ f[6] = 0.85,\ f[7] = 0.95,\ f[8] = 1.0.$

Users select one of them.

5.3 PREDICTING INVESTMENT RESULTS

Two investment possibilities are considered: Certificates of Deposit (CD) and shares. Considering CD, the two parameters should be fixed: the interest and the reliability (bank's survival probability). These parameters are defined in the file *Portfolio.java*. Considering shares, only

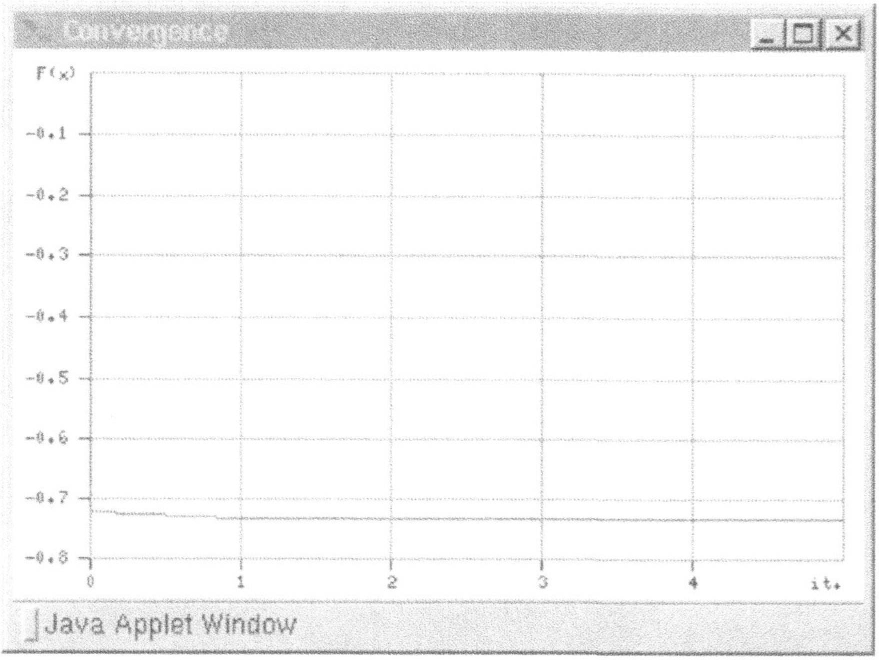

Figure 13.4. The convergence window.

one parameter is used, the predicted stock rate at the time of CD maturity. That is correct, if the investor is confident[4] about the predicted stock rates at the maturity time. The default values of expected changes of stock rates are zero. The predicted stock rates are given in the files like this one: *onestep?VB.txt.*

```
Input file = VB.txt
Pradzia = 26.000000 Prognoze = 30.297932 </BODY></HTML>
```

Here 'VB' denotes the Vilnius Bank stock rates. 'Pradzia' denotes the rates now. 'Prognoze' denotes the predicted rates. The difference between predicted and present stock rates, divided by the present stock rate and expressed in percents, is considered as a "stock interest." That is an important simplification. However, it helps to consider both shares and CD in the same way. That is convenient for calculations.

[4]Without the "confidence" assumption one should consider stock rates predictions errors, too.

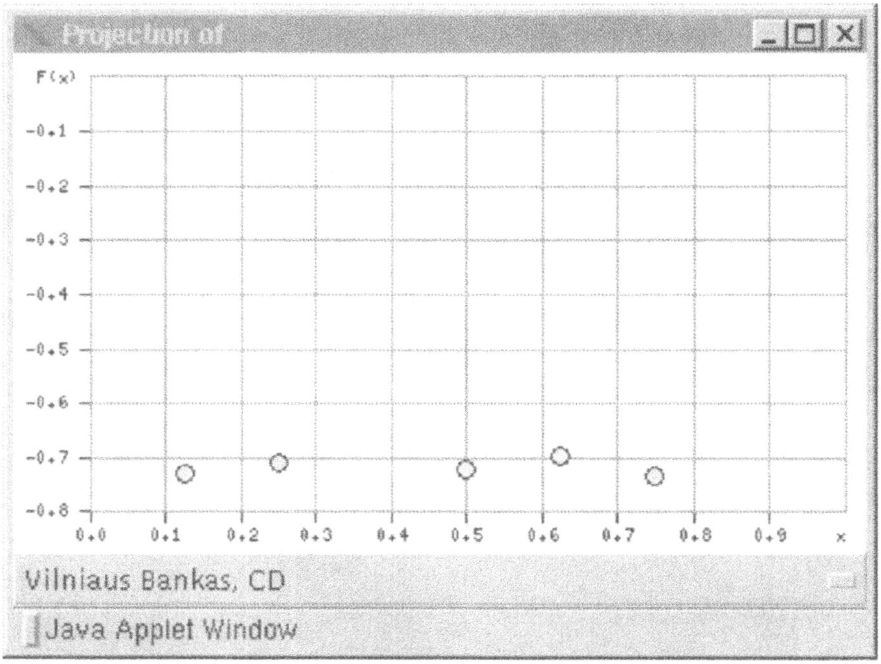

Figure 13.5. The projection window.

Predicted rates in *VB.txt* files are obtained by the ARMA model using the Vilnius Bank data. The predicted values can be changed manually by clicking and writing on the corresponding fields of the 'Tasks' page.

5.4 DATA

The stock rates of the following major Lithuanian joint-stock companies, including the banks, are considered.

Bank "Vilniaus Bankas" : data file VB.txt,

Bank "Snoras" : data file Snoras.txt,

Bank "Hermis" : data file Hermis.txt,

Savings bank "LTB" : data file LTB.txt,

Agricultural bank "LZUB" : data file LZUB.txt,

Bank "Siauliu Bankas" : data file Siauliu.txt,

Economic bank "Ukio Bankas" data file Ukio.txt,

Dairy "Rokiskio Suris" : data file Rokiskio.txt,

Brewery "Kalnapilis" data file Kalnapilis.txt,

Dairy "Birzu Pienas" : data file BirzuAB.txt.

The first record of the data files[5] shows closing rates on January 01, 1998. The last record defines closing rates on November 14, 1998. Two hundred twenty nine sessions of Lithuanian stock exchange are recorded in the data. Using this data ARMA model makes ninty day predictions what corresponds to three month CD.

5.5 RESULTS

Optimization results are in Tables 13.1, 13.2, and 13.3. Tables 13.1 and 13.2 both are for a prudent person. Table 13.1 is obtained by the Monte Carlo method. Table 13.2 shows results of the deterministic method. Table 13.3 is obtained by the deterministic method for a risky person.

The "Optimal investment" column is just a copy of the output field (see Figure 13.3). The output field represents optimal values of internal variables. The internal variables are proportional to the optimal parts of the capital. Actual optimal parts are obtained by normalization [6] and are shown in the "Optimal part" column.

Tables 13.1 and 13.2 show that the optimal distribution of capital obtained by the approximate method differ from that defined by the exact one. This is important disadvantage of the approximate Monte Carlo method. Note, that the best values of utility functions are close for both methods. They are 0.773, for the Monte Carlo method, and 0.776, for the deterministic one. This means that the optimal values of utility functions are not very sensitive to moderate changes in the capital distribution. The reason is that here investment alternatives are not too different.

[5] The data file is on the web-sites (see Section 4.).

[6] Users make the normalization in this software version.

Company	Optimal investment	Optimal part
Vilniaus Bankas, CD	0.157	0.0251
Snoras, CD	0.354	0.0566
Hermis, CD	0.402	0.0643
Litimpex, CD	0.724	0,1159
Ukio Bankas, CD	0.334	0.0535
Siauliu Bankas, CD	0.842	0.1348
Medicinos Bankas, CD	0.096	0.0153
State bonds	0.318	0.0509
Vilniaus Bankas, shares	0.626	0.1002
Snoras, shares	0.677	0.1083
Hermis, shares	0.293	0.0469
Ukio Bankas, shares	0.466	0.0746
Siauliu Bankas, shares	0.461	0.0738
Rokiskio Suris, shares	0.108	0.0169
Birzu Pienas, shares	0.061	0.0097
Kalnapilis, shares	0.146	0.02337

Table 13.1. Monte Carlo method. Prudent person. Optimal portfolio obtained after 892 iterations. The utility is 0.773.

Comparing Table 13.2 representing the prudent person and Table 13.3 representing the risky one, we see differences. The best value of utility function is equal to 0.776, for the "prudent." It is equal to 0.291, for the "risky." This difference is explained by assumptions made to normalize utility functions. The normalization assumes that the maximal utility is a unit and the minimal one is zero. Therefore, the best value of utility function of the risky person happens to be less then that of the prudent one. However, lower utilities does not mean that risky persons are getting less satisfaction. One needs more sophisticated normalization techniques to compare correctly the satisfaction scales of different persons by their utility functions.

5.6 FUTURE DEVELOPMENTS

The first step is to drop the "confidence" assumption that one predicts the stock rates exactly. To do this, the variance of the predicted stock rates should be included into the updated Portfolio Model.

The important future task is the sensibility analysis. For example, by considering three scenarios: pessimistic, optimistic and realistic. In

Company	Optimal investment	Optimal part
Vilniaus Bankas, CD	0.699	0.0755
Snoras, CD	0.453	0.0489
Hermis, CD	0.836	0.904
Litimpex, CD	0.795	0,0859
Ukio Bankas, CD	0.366	0.0395
Siauliu Bankas, CD	0.783	0.0846
Medicinos Bankas, CD	0.926	0.1001
State bonds	0.589	0.0636
Vilniaus Bankas, shares	0.937	0.1001
Snoras, shares	0.361	0.0390
Hermis, shares	0.343	0.0370
Ukio Bankas, shares	0.002	0.0002
Siauliu Bankas, shares	0.944	0.1020
Rokiskio Suris, shares	0.760	0.0822
Birzu Pienas, shares	0.07	0.00075
Kalnapilis, shares	0.052	0.0056

Table 13.2. Exact o method. Prudent person. Optimal portfolio obtained after 892 iterations. The utility is 0.776.

the pessimistic scenario lower probabilities p_i for CD's and lower interest rates for stocks are considered. In the optimistic case one sets higher values for these parameters. The realistic case is in the middle. Important task is to represent visually the "stability" of securities [7]. If the optimal values of some securities are not too different in all three scenarios, they are considered as stable.

[7] CD and Stocks.

Company	Optimal investment	Optimal part
Vilniaus Bankas, CD	0.907	0.107
Snoras, CD	0.083	0.0098
Hermis, CD	0.857	0.101
Litimpex, CD	0.438	0,0517
Ukio Bankas, CD	0.024	0.00283
Siauliu Bankas, CD	0.737	0.0870
Medicinos Bankas, CD	0.739	0.872
State bonds	0.656	0.0774
Vilniaus Bankas, shares	0.762	0.0899
Snoras, shares	0.179	0.0211
Hermis, shares	0.894	0.1055
Ukio Bankas, shares	0.451	0.0532
Siauliu Bankas, shares	0.098	0.011
Rokiskio Suris, shares	0.946	0.112
Birzu Pienas, shares	0.019	0.00224
Kalnapilis, shares	0.076	0.0089

Table 13.3. Exact method. Risky person. Optimal investment obtained after 159 iterations. The utility is 0.291.

Chapter 14

EXCHANGE RATE PREDICTION, TIME SERIES MODEL

1. INTRODUCTION

Solutions of optimal investment problems, considered in the previous chapter, depend on predicted stock rates. Predictions are needed considering most of the investment problems, because the result of an investment depends on future values of various parameters. Predictions are an important part of scheduling problems, too. The optimal schedules depend on the predicted demand, and supply. Time series models are common prediction tools. Here we investigate a well-known time series model. The so called, Auto Regressive Moving Average (ARMA) model. We briefly consider the Artificial Neural Network (ANN) and the Bilinear models, too.

Mainly, the financial data is predicted. The prediction of call rates, using the same models, are briefly mentioned, just for comparison. The call rate prediction is described in detail considering call center optimization problems in Chapter 15. There the traditional ARMA is supplemented by an expert model.

Modeling economic and financial time series by ARMA has attracted the attention of many researchers in recent years (Diebold and Rudebusch, 1989; Cheung, 1993; Yin-Wong and Lai., 1993; Cheung and Lai, 1993; Koop et al., 1994; Mockus and Soofi, 1995). Three approaches have been used to estimate the parameters of the ARMA models, : Maximum Likelihood (ML) (Sowel, 1992), approximate ML (Li and McLeod,

1986; Fox and Taqqu, 1986; Hosking, 1981; Hosking, 1984), and two-step procedures (Geweke and Porter-Hudak, 1983; Janacek, 1982). In all the cases local optimization techniques were applied. Here the optimization results depend on the initial values, what implies that one cannot be sure if the global maximum is found.

The global optimization is very difficult in many cases[1]. The reason is a high complexity of multi-modal optimization problems. It is well known (Ko, 1991) that optimization of polynomial-time computable real functions cannot be done in polynomial-time, unless $P = NP$[2]. In practice, this means that we need an algorithm of exponential time to obtain the ϵ-exact solution. The number of operations in exponential algorithms grows exponentially with the accuracy m of solution and dimension n of the optimization problem. The accuracy m means that $\epsilon \leq 2^{-m}$. The dimension n means that one optimizes a function $f(x)$ where $x = (x_1, ..., x_n)$.

The Least Squares (LS) method is a popular approach to estimate parameters of ARMA models. Using LS one minimizes the log-sum of square residuals using ARMA models and their extensions (Mockus and Soofi, 1995). In this chapter, the multi-modality problems are considered using different data. The data include:
daily exchange rates of \$/£ and DM/\$,
closing rates of stocks of AT&T , Intel Co, and some Lithuanian banks, the London stock exchange index (Raudys and Mockus, 1999),
call-rates of a commercial call-center.
The graphical images and comparison of the average prediction results of ARMA and the Random Walk (RW) models are presented.

[1]By the term 'difficult' we mean the time measure of computational complexity, that is, the length of time would be needed for a standard universal computer to do a task.
[2]The notation $P = NP$ means the existence a polynomial-time algorithm P for solving NP-complete problems. That is merely a theoretical possibility.

2. AUTO REGRESSIVE MOVING-AVERAGE MODELS (ARMA)

2.1 DEFINITIONS

We define ARMA model as

$$w_t = \sum_{i=1}^{p} a_i w_{t-i} + \sum_{i=1}^{q} b_i \epsilon_{t-i} + \epsilon_t. \tag{14.1}$$

We assume that

$$z_{t-i} = 0,\ w_{t-i} = 0,\ \epsilon_{t-i} = 0,\ \ if\ \ t \leq i. \tag{14.2}$$

2.2 DEFINITION OF RESIDUALS

An advantage of residual minimization is that one may see directly how the objective depends on unknown parameters. Using equalities (14.1) we define residuals by recurrent expressions

$$\begin{aligned} \epsilon_1 &= w_1, \qquad (14.3) \\ \epsilon_2 &= w_2 - a_1 w_1 - b_1 \epsilon_1, \\ &\ldots\ldots\ldots\ldots\ldots\ldots\ldots\ldots \\ \epsilon_t &= w_t - a_1 w_{t-1} - \ldots - a_p w_{t-p} - b_1 \epsilon_{t-1} - \ldots - b_q \epsilon_{t-q}. \end{aligned} \tag{14.4}$$

Next the sum

$$f(x) = \log f_m(x),\ \ f_m(x) = \sum_{t=1}^{T} \epsilon_t^2 \tag{14.5}$$

is minimized. The logarithm is there to decrease the objective variation by improving the scales.

3. MINIMIZATION OF RESIDUALS OF ARMA MODELS

Here we consider an algorithm for optimization of parameters of the ARMA model. This model is simple and may be considered as a good first approximation. Denote by y_t the value of y at the time t. Denote by $a = (a_1, ..., a_q)$ a vector of auto-regression (AR) parameters, and by $b = (b_1, ..., b_q)$ a vector of moving-average (MA) parameters.

$$y_t - \sum_{i=1}^{p} a_i y_{t-i} = \epsilon_t - \sum_{j=1}^{q} b_j \epsilon_{t-j}, \ t = 1, ..., T. \tag{14.6}$$

The residual

$$\epsilon_t = y_t - \sum_{i=1}^{p} a_i y_{t-i} + \sum_{j=1}^{q} b_j \epsilon_{t-j}, \tag{14.7}$$

or

$$\epsilon_t = B_t + \sum_{i=1}^{p} a_i A_t(i). \tag{14.8}$$

Here

$$B_t = y_t + \sum_{j=1}^{q} b_j B_{t-j-1}, \tag{14.9}$$

and

$$A_t(i) = -y_{t-i-1} + \sum_{j=1}^{q} b_j A_{t-j-1}, \tag{14.10}$$

where $t - i > 0$ and $t - j > 0$.

3.1 OPTIMIZATION OF AR PARAMETERS

Denote

$$S(a, b) = \sum_{t=1}^{T} \epsilon^2, \tag{14.11}$$

where $a = (a_1, ..., a_p)$ and $b = (b_1, ..., b_q)$. From expressions (14.11) and (14.8) the minimum condition is

$$\frac{\partial S(a, b)}{\partial a_j} = 2 \sum_{t=1}^{T} \epsilon_t A_t(j) = 0, \ j = 1, ..., p, \tag{14.12}$$

or

$$\sum_{i=1}^{p} A(i, j) a_i = -B(j), \ j = 1, ..., p. \tag{14.13}$$

Here

$$A(i,j) = \sum_{t=1}^{T} A_t(i)A_t(j), \tag{14.14}$$

and

$$B(j) = \sum_{t=1}^{T} A_t(j)B_t. \tag{14.15}$$

The minimum of expression (14.11) at fixed parameters b is defined by the system of linear equations

$$a(b) = A^{-1}B. \tag{14.16}$$

Here matrix $A = (A(i,j),\ i,j = 1,...p)$. Vector $B = (B(j),\ j = 1,...p)$, where elements $A(i,j)$ are from (14.14), components $B(j)$ are from (14.15), and A^{-1} is an inverse matrix A. This way one define the vector $a(b) = (a_i(b),\ i = 1,...,p)$ that minimize sum (14.11) at fixed parameters b.

3.2 OPTIMIZATION OF MA PARAMETERS

The sum of squared residuals (14.11) is a non-linear non-convex function of parameters b. Thus, we have to consider the global optimization algorithms. Denote

$$f(x) = \log S(a(x), x). \tag{14.17}$$

Here $x = b$ and $S(a,b)$ is from (14.11) at optimal parameter $a = a(b)$. Denote

$$b^0 = x^0 = arg \min_x f(x). \tag{14.18}$$

3.3 PREDICTING "NEXT-DAY" RATE

Predicting "Next-Day" Rate, one minimizes the expected "next-day" squared deviation ϵ_{t+1} using the data available at the moment t

$$y^0_{t+1} = arg \min_{y_{t+1}} \mathbf{E}\ \epsilon^2_{t+1}. \tag{14.19}$$

Here

$$\mathbf{E}\epsilon^2_{t+1} = \mathbf{E}\ (B_{t+1} + \sum_{i=1}^{p} A_{t+1}(i)a_i(b^0))^2, \tag{14.20}$$

where the optimal parameter b^0 was obtained using the data available at the day t. Variance (14.20) is minimal, if

$$y_{t+1}^0 = B_{t+1} + \sum_{i=1}^{p} A_{t+1}(i) a_i(b^0), \tag{14.21}$$

because the expectation of y_{t+1} is $B_{t+1} + \sum_{i=1}^{p} A_{t+1}(i) a_i(b^0)$ under the assumptions.

3.4 EVALUATION OF ARMA PREDICTION ERRORS

We compare the "next-day" ARMA prediction results and a simple Random Walk (RW) model. In RW model the next day value is $y_{t+1} = y_t + \epsilon_{t+1}$ and the predicted value is equal to the conditional expectation of $y_{t+1} = y_t$. This means that the present value is the RW prediction for the next day. There are three reasons to consider RW models

- simplicity of RW,
- RW is a trivial case of ARMA where $q = 0$, $p = 1$, and $a_1 = 1$,
- RW represents unpredictable time series, because random numbers ϵ_{t+1} are independent.

Therefore, the difference of average prediction errors of ARMA and RW models can be regarded as some parameter of "ARMA-unpredictability" of the data. That means that if this difference is zero or negative, then the time series are not predictable by the ARMA model. Table 14.1 shows the difference between the mean square deviations of ARMA and RW models. The table uses four data sets

- exchange rates of DM/\$, \$/£, Yen/\$, and Fr/\$,
- closing rates of AT&T , Intel Co., and Hermis Bank stocks ,
- index of the London Stock Exchange,
- call rates of a call center.

In Table 14.1, the symbol $MeanARMA$ denotes average prediction errors of ARMA model. The symbol $VarArma$ means the variance of

Table 14.1. The average "next-day" prediction results of ARMA and RW models.

Data	*DeltaARMA%*	*MeanARMA*	*VarARMA*
\$/£)	- 1.779090e-01	1.293609e+00	8.454827e-02
DM/\$	- 1.191e-02	1.092e-01	9.985e-02
Yen/\$	- 1.086e+00	6.369e+00	6.446505e+00
Fr/\$	- 3.029e-01	4.285e-01	3.395e-01
AT&T	-1.375e+00	4.554e+00	3.621e+00
Intel Co	+2.814e-01	2.052e+01	1.936e+00
Hermis Bank	-4.280e+01	2.374e+01	1.998e+01
London Stock Exchange	-5.107e-01	2.751e+02	2.346e+01
Call Center	+3.076e+01	8.453e+02	7.111e+02

ARMA predictions. The symbol $DeltaARMA\%$ denotes the relation $(RW - ARMA)/RW$ in percentages. Herte RW defines average errors of the Random Walk. $DeltaARMA\% > 0$ means that the ARMA model predicts better.

The data is divided into three equal parts.
The first part is to estimate parameters a and b of an ARMA model using fixed numbers p and q.
The best values of p and q are defined using the second part of data.
The third part is to compare ARMA and RW models. The table shows the comparison results.

Table 14.1 shows that the ARMA model predicts all the financial data not better than RW. However, ARMA predicts call rates thirty one percent better then RW. That is a statistically significant difference. The observed deviations between RW and ARMA models of financial data are too small for practical conclusions. Figure 14.1 shows optimal parameters b and a. Here the optimal $q = 1$ and $p = 12$. The "multi-day" predictions of call rates are shown in Figure 14.2.

4. EXTERNAL FACTORS

Often we have to predict one main factor that depends on some external factors. This means that we are not interested in future values of external factors while predicting the main one. For example, we predict the stock rate as the main factor depending on the GNP as the external

```
 b[0] =   -2.888616e-01
a[0] = 9.131314e+02
a[1] = 4.565372e-02
a[2] = 3.765721e+02
a[3] = 1.129722e-01
a[4] = 2.993138e+02
a[5] = 3.935438e-02
a[6] = 1.109620e+02
a[7] = 2.491347e-02
a[8] = 3.427414e+01
a[9] = -1.073026e-01
a[10] = -1.432491e+03
a[11] = 8.030343e-01
```

Figure 14.1. The optimal parameters of the ARMA model predicting call rates.

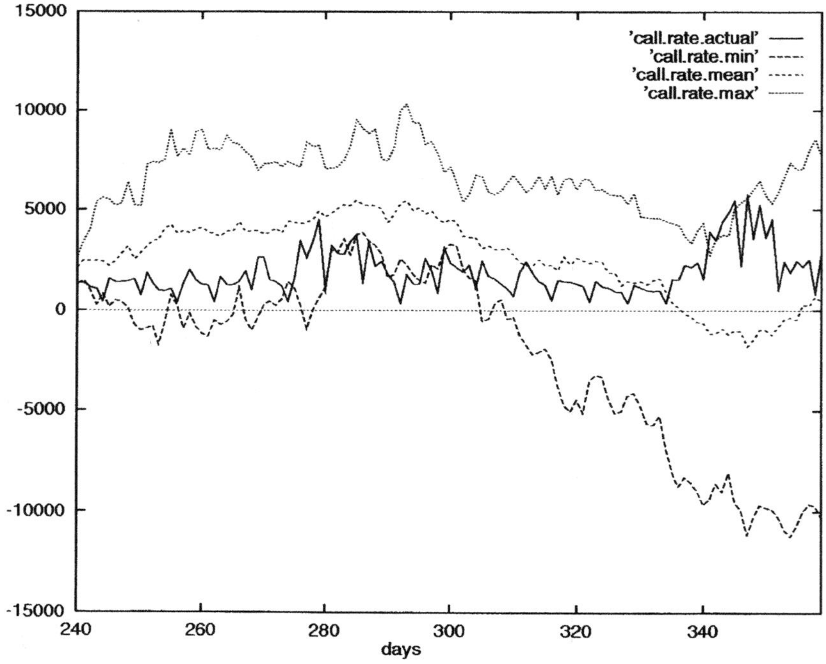

Figure 14.2. Multi-day predictions of call rates.

factor. We consider the cases when future values of all the factors are not known. Considering the cases when future values of external factors are known[3] the model should be modified accordingly.

Different symbols are used in ARMA models with external factors. The predicted value is denoted by $\nu(t)$ and the external factors are $\eta(t) = (\eta_1(t), \eta_2(t), ...)$. An influence of some external factors may be delayed.

For illustration, consider two-dimensional case, omitting the Moving Average (MA) part. This means the two-dimensional Auto Regressive (AR) model with an external factor $\eta(t)$

$$\nu(t) = \sum_i^p (a_{1i}\nu(t-i) + a_{2i}\eta(t-d-i)) + \epsilon(t), \; p+d < t < T_0. \tag{14.22}$$

Here d is the delay parameter. First we minimize the squared deviation $\sum_t \epsilon(t)^2$ at fixed delay d and then chose the optimal delay d. The minimum of a sum (14.22), as a function of parameters a, is defined by the system of linear equations. These are obtained from the condition that all the partial derivatives are zero

$$\sum_{t=p+d+1}^{T_0-1} (\nu(t) - \sum_i^p (a_{1i}\nu(t-i) + a_{2i}\eta(t-d-i))\nu(t-i) = 0, \; i = 1, ..., p. \tag{14.23}$$

$$\sum_{t=p+d+1}^{T_0-1} (\nu(t) - \sum_i^p (a_{1i}\nu(t-i) + a_{2i}\eta(t-d-i))\eta(t-d-i) = 0, \; i = 1, ..., p. \tag{14.24}$$

We have to solve $2p$ linear equations with $2p$ variables $a_{1i}, a_{2i}, \; i = 1, ..., p$ and obtain the least squares estimates $a_{1i}(d), a_{2i}(d), \; i = 1, ..., p$ at given d.

Sum (14.25), as a function of delay parameter d, is not necessarily uni-modal one. Thus to define the exact minimum

$$\nu(t) = \sum_i^p (a_{1i}(d)\nu(t-i) + a_{2i}(d)\eta(t-d-i)) + \epsilon(t), \tag{14.25}$$

$$p + d < t < T_0.$$

[3]For example, while predicting call rates we know in advance such "external" factors as holidays, vacation times, e.t.c.

One should consider all the $K(d)$ values of the integer d. Here $K(d)$ is the number of "interesting" values of d. Therefore, expression (14.25) is an example of multi-modal function in AR models. We mentioned the delay time just to show a way how to include this important factor into a time series model. Later the delay factor is omitted, to simplify the expressions.

4.1 MISSING DATA

Suppose that predicting $\nu(t)$ we don't know some previous values $\nu(s),\ s < t$. Then we replace the missing data by the expected value $\hat{\nu(s)}$ defined as

$$\hat{\nu(s)} = \sum_{i}^{p}(a_{1i}\nu(s-i) + a_{2i}\eta(s-i)),\ p < t < T_0. \tag{14.26}$$

The least squares estimates of the regression parameters $a_{1i},\ a_{2i}$ are obtained using observations before the missing one and minimizing expression (14.22). The same idea can be extended, if two or more values of the factor ν are missing.

This algorithm is to "fill" the missing data only for the predicted factor ν. We replace missing values of the external factor η by the nearest previous value, because we do not predict the external factors in this model.

5. APPLYING ARMA TO EXTERNAL FACTORS

Omitting the delay d in expression (14.22) we write

$$\nu(t) = \sum_{i}^{p}(a_{1i}\nu(t-i) + a_{2i}\eta(t-i)) + \epsilon(t),\ p < t < T_0. \tag{14.27}$$

Including this expression into the ARMA model (14.6) one obtains the following expression

$$\nu(t) = \sum_{i}^{p}(a_{1i}\nu(t-i) + a_{2i}\eta(t-i)) + \sum_{i=1}^{q} b_i\epsilon_{t-i} + \epsilon_t,\ p < t < T_0. \tag{14.28}$$

Extending expression (14.28) to M external factors

$$w^M(t) = \sum_{m=1}^{M} \sum_{i}^{p} a_i^m w^m(t-i) + \sum_{i=1}^{q} b_i \epsilon_{t-i} + \epsilon_t, \; p < t < T_0. \tag{14.29}$$

In expression (14.29), the predicted and the external factors are denoted by the same letter w with different upper indices. All the factors are represented as one-dimensional time series of the special type

$$w(Mt) = \sum_{m=1}^{M} \sum_{i}^{p} \alpha_{Mi-(M-m)} w(M(t-i)-(M-m)) + \sum_{i=1}^{q} b_i \epsilon_{M(t-i)} + \epsilon_{Mt}, \; p < t < T_0. \tag{14.30}$$

Here $\alpha_{Mi-(M-m)} = a_i^m$, $w(M(t-i)-(M-m)) = w^m(t)$, $m = 1, ..., M$ and the index $mi = m * i$. Using this expression one applies the software developed for the one-dimensional ARMA model to M external factors. The data file should correspond to the expression (14.30).

6. ARTIFICIAL NEURAL NETWORKS MODELS (ANN)

With respect to the Moving Average parameters b, the ARMA model is non-linear (see expression (14.29)). With respect to data, this model is linear. If we are interested in the non-linearities, then we may apply many other non-linear models, including the ones that are non-linear with respect to data. In this book, we discuss two of them. Here the ANN model will be considered. In the next chapter, we introduce a bilinear term into the ARMA model.

6.1 INVOLVING AUTO REGRESSION (AR) INTO ANN

We apply ANN by involving the non-linear activation function ϕ into the standard Auto-Regression (AR) model

$$w_t = \phi(\sum_{i=1}^{p} a_i w_{t-i}) + \epsilon_t. \tag{14.31}$$

The idea lurking behind ANN-AR model is that the activation function ϕ roughly represents the activation of a real neuron. We minimize the sum

$$f_m(x) = \sum_{t=1}^{T} \epsilon_t^2. \tag{14.32}$$

Here the objective $f_m(x)$ depends on l unknown parameters given as a l-dimensional vector $x = (x_k, k = 1, ..., p) = (a_i, i = 1, ..., p)$.

One see from expression (14.31) that residuals ϵ_t are non-linear functions of parameters a_t[4]. This means that the minimum conditions

$$\frac{\partial f_m(x)}{\partial a_i} = 0, \ i = 1, ..., p \tag{14.33}$$

is a system of non-linear equations with multiple solutions.

An interesting activation function is derived using the Gaussian distribution function

$$\phi(w_t(l)) = \frac{\beta}{\sqrt{2\phi\sigma}} \int_{-\infty}^{w_t(l)} e^{-\frac{w-\mu}{\sigma}} dw. \tag{14.34}$$

Here $w_t(l) = \sum_{i=1}^{l} a_i w_{t-i}$. β is a scale parameter. The function (14.34) is different from the activation of a real neuron[5] but is convenient for analysis. Here sum (14.31) depends on the parameters β, σ, μ, too. These parameters are unknown, as usual. Therefore, they should be optimized. Therefore, the objective $f_m(x)$ depends on $p + 3$ unknown parameters represented as a $p + 3$-dimensional vector $x = (x_k, k = 1, ..., p + 3) = (a_i, i = 1, ..., p, \beta, \sigma, \mu)$. That is the main difference of model (14.34) from the traditional ANN models.

In the traditional ANN models the activation functions are selected by their resemblance to the natural ones from biophysical experimentation. In this research the resemblance factor is neglected. The activation function (14.34) is considered just as a reasonable non-linearity that should be adapted to the available data. The multi-modality problems of ANN models are discussed in (Mockus et al., 1997).

---

[4]This is true, if the activation function ϕ is non-linear. The linear activation function reduces the ANN model to the standard Auto Regressive (AR)n model.

[5]The output of the traditional activation function is supposed to be non-negative.

7. BILINEAR MODELS (BL)

It is well known that for the adequate description of some phenomena additional non-linear terms of the time series should be included. A simple example is a bilinear term (see (Rao and Gabr, 1984; Liu, 1989)). Here bilinear time series extend the ARMA model:

$$w_t = \sum_{i=1}^{p} a_i w_{t-i} + \sum_{i=1}^{q} b_i \epsilon_{t-i} + \sum_{i=1}^{s} \sum_{j=1}^{r} c_{ij} z_{t-i} \epsilon_{t-j} + \epsilon_t. \tag{14.35}$$

An illustrative example is in (Mockus et al., 1997).

8. AUTO REGRESSIVE FRACTIONALLY INTEGRATED MOVING AVERAGE MODELS (ARFIMA)

8.1 DEFINITIONS

ARMA, ANN, and BL models consider stationary time series . The stationarity of a model is a simplification of reality. A well-known source of non-stationary behavior is the linear component, the trend. One eliminates the trend by differentiation, since derivatives of linear functions are constant. The elegant extension of this idea is the Auto Regressive Fractionally Integrated Moving Average model (ARFIMA).

We define an ARFIMA[6] process as the following time series z_t[7]

$$A(L)(1-L)^d z_t = B(L)\epsilon_t. \tag{14.36}$$

Here

$$A(L)w_t = w_t - \sum_{i=1}^{p} a_i w_{t-i} \tag{14.37}$$

and

$$B(L)\epsilon_t = \epsilon_t - \sum_{i=1}^{q} b_i \epsilon_{t-i}, \tag{14.38}$$

[6]Often the alternative title ARIMA(p,d,q) is used.

[7]Note that, contrary to the traditional practice, we do not assume the time series to be stationary. We allow the degree of integration of the series to be detected by direct estimate of the differentiation parameter d.

where $\epsilon_t = Gaussian \ \{0, \sigma^2\}$. We define the transformation $(1-L)^d$ as follows:

$$w_t = (1-L)^d z_t = z_t - \sum_{i=1}^{\infty} d_i z_{t-i}. \tag{14.39}$$

Here

$$d_i = \frac{\Gamma(i-d)}{\Gamma(i+1)\Gamma(-d)}, \tag{14.40}$$

where d is a fractional integration parameter, and $\Gamma(.)$ is a gamma function. We assume that

$$z_{t-i} = 0, \ w_{t-i} = 0, \ \epsilon_{t-i} = 0, \quad if \ \ t \leq i. \tag{14.41}$$

We truncate sequence (14.39)

$$d_i = 0, \quad if \ \ i > R. \tag{14.42}$$

Here R is the truncation parameter, the number of non-zero components.

8.2 MINIMIZATION OF RESIDUALS

We define residuals by recurrent expressions:

$$\begin{aligned} \epsilon_1 &= w_1, & (14.43) \\ \epsilon_2 &= w_2 - a_1 w_1 + b_1 \epsilon_1, \\ & \ldots\ldots\ldots\ldots\ldots\ldots\ldots\ldots\ldots\ldots \\ \epsilon_t &= w_t - a_1 w_{t-1} - \ldots - a_p w_{t-p} + b_1 \epsilon_{t-1} + \ldots + b_q \epsilon_{t-q}. \\ & & (14.44) \end{aligned}$$

Next, the sum

$$f(x) = \log f_m(x), \quad f_m(x) = \sum_{t=1}^{T} \epsilon_t^2 \tag{14.45}$$

is minimized.

The logarithm is used to decrease the objective variation by improving the scales. The objective $f_m(x)$ depends on $m = p + q + 1$ unknown parameters. They are represented as an m-dimensional vector $x = (x_k, k = 1, ..., m) = (a_i, i = 1, ..., p, b_j, j = 1, ..., q, d)$.

It follows from (14.44), (14.39), and (14.37) that residuals ϵ_t are linear functions of the parameters a_t. This means that the minimum conditions

$$\frac{\partial f_m(x)}{\partial a_i} = 0, \ i = 1, ..., p \tag{14.46}$$

are given by a system of linear equations. They estimate linear parameters $a_i = a_i(b, d)$ as a function of non-linear ones $b_i, i = 1, ..., q, d$. It reduces the number of parameters of non-linear optimization to $n = q+1$.

The system

$$\frac{\partial f_m(x)}{\partial b_i} = 0, \ i = 1, ..., q \tag{14.47}$$

may have a multiple solution, because the residuals ϵ_t depend on b_i as polynomials of degree $T - 1$.

The equation

$$\frac{\partial f_m(x)}{\partial d} = 0 \tag{14.48}$$

may have multiple solutions, too, because the residuals depend on d as a polynomial of degree R, where R is a truncation parameter.

The objective $f_m(x)$ is a multi modal function of parameters d and $b_i, i = 1, ..., q$ [8]. Therefore, one uses methods of global optimization (see, (Mockus et al., 1997)). Denote

$$f(x) = \log f_{q+1}(x). \tag{14.49}$$

Here

$$f_{q+1}(x) = f_m(x), \ x_j = b_j, j = 1, ..., q, \ x_{j+1} = d,$$
$$x_{j+1+i} = a_i(b, d), \ i = 1, ..., p.$$

This means that condition (14.46) defines those x- components that represent parameters $a_i, i = 1, ..., p$.

There is no variance σ^2 in expressions (14.49) and (14.44). If necessary, we have to estimate the variance by another well-known technique.

[8] The same reasoning applies to the log-likelihood function, too.

8.3 DISCUSSIONS

Table 14.2 shows results obtained using the ARFIMA model (Mockus et al., 1997). Parameters b and d were estimated using daily exchange rates of \$/£, and DM/\$, and closing rates of AT&T and Intel Co shares. Table 14.2 shows that d's are very close to zero (see also Figures 14.3).

Table 14.2. Estimated parameters b and d of ARFIMA models.

Data	b_0	b_1	d	$\min \log f(x)$
\$/£	-1.195	-0.169	0.0005	1.51675
DM/\$	-1.019	0.0120	0.0007	1.60065
AT&T	-1.017	0.0118	0.00005	9.83208
Intel Co	0.9975	0.0055	0.012	7.35681

An exeption is Intel Co shares.

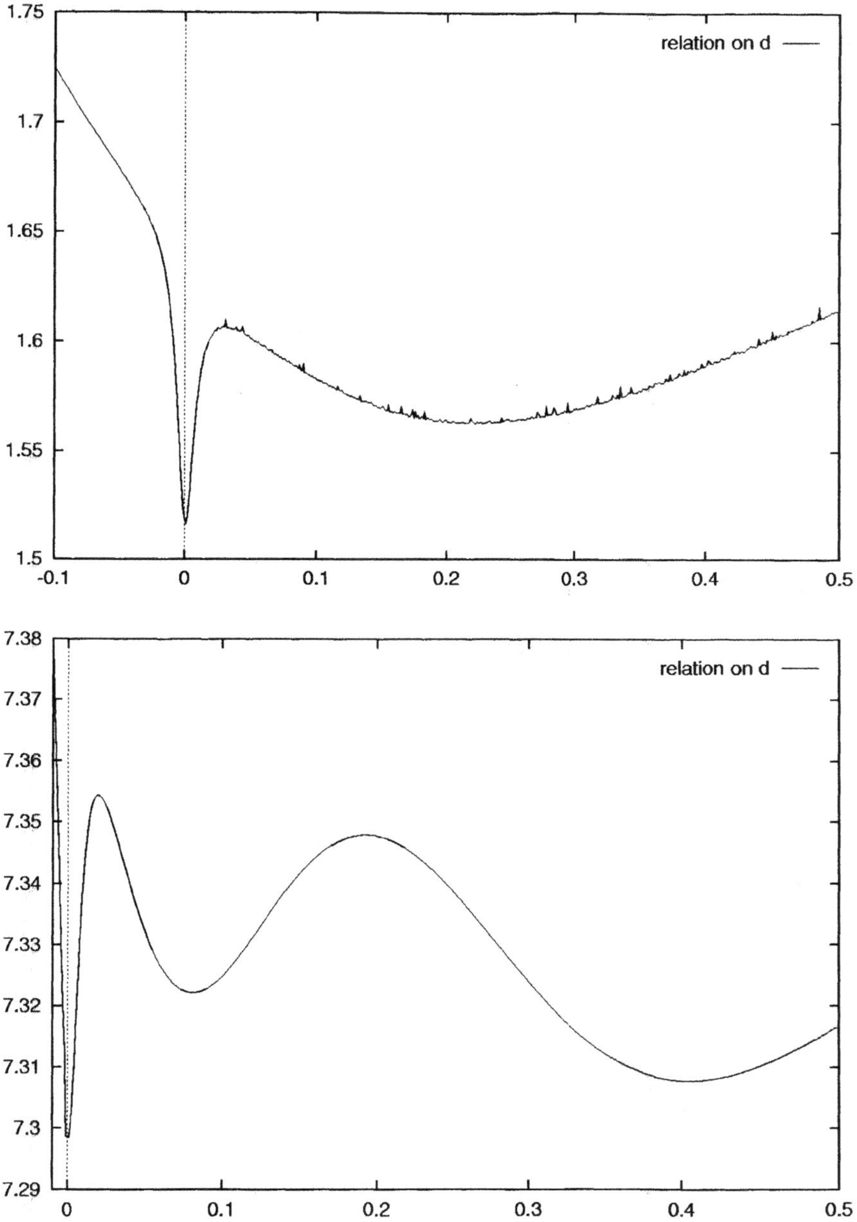

Figure 14.3. Log-sum (14.45) as a function of the parameter $d \in [-0.01, 0.5]$. Top figure shows \$/£ exchange rates. Bottom one shows Intel Co stocks closing rates.

Hence, following the traditional approaches (Cheung, 1993; Yin-Wong and Lai., 1993; Cheung and Lai, 1993) one concludes that the underlying stochastic processes generating exchange rates do not exhibit persistence and are stationary. That contradicts the visual impression of the corresponding data (see Figures 14.4 and 14.10).

This apparent contradiction may be resolved by dropping the assumption that the parameters a, b, d of the ARFIMA model remain constant. This assumption is common for most of the traditional methods. An alternative is the structural stabilization model described in Chapter 10..

9. MULTI-STEP PREDICTIONS

Often one wants to predict for several days (hours, weeks, months, e.t.c) ahead. This is the Multi-Step Predictions (MSP). One can do that by using a Monte Carlo simulation. The residuals ϵ_t (see expression (14.44)) are determined up to the simulation starting moment $t(s)$ using the observed data. The rest of residuals $\epsilon_t,\ t \geq t(s)$ are generated by a Gaussian distribution with zero mean and variance σ^2. The simulation is repeated K times. Considering external factors one just replaces them by the nearest previous values (see Section 4.1).

The illustration is in Figure 14.2. The line *call.rate.actual* shows the observed call rate. Lines
call.rate.min, *call.rate.mean*, and *call.rate.max* show the minimal, the average, and the maximal results of MSP predictions. The "min" and "max" lines denote the lower and the upper values of simulation. Therefore, these lines are called "MSP- confidence intervals," meaning that if the model is true, one may expect those "intervals" to cover the real data with some "MSP-confidence level" $\alpha(MSP)$.

It is difficult to define $\alpha(MSP)$ exactly. If "interval deviations" may be considered as independent and uniformly distributed random variables, we obtain $\alpha(MSP) = 1 - K$. Here K is the number of Monte-Carlo repetitions. In the example, $K = 10$, thus $\alpha(MSP) = 0.9$.

This assumption over-simplifies the statistical model. Therefore, the "MSP-confidence level" $\alpha(MSP)$ is just a Monte Carlo approximation.

10. STRUCTURAL STABILIZATION

10.1 STABILIZATION OF STRUCTURES OF TIME SERIES

The traditional time series models assume that their parameters do not change. Examples are parameters a, b of the ARMA model, parameters a, b, d of the ARFIMA model, parameters a, b, c of the BL model, and parameters a, β, σ, μ of the ANN model. Natural changes of economical conditions introduce a variability of parameters of models describing the financial data, such as stock rates or currency exchange rates. The variability remains in the stable economical conditions, too. The reason is the "Feed-Back" processes. It is well known that the stock rate predictions may influence the supply and demand and consequently the future rates. In such cases the statistical best fit models should be complemented by the game-theoretical equilibrium models. There is the possibility that the equilibrium model would be namely the Wiener process. That means that market rates in economics are governed by models similar to the Brownian motion. In such a case market rates would be as unpredictable as movements of individual molecules in gases. This is a working hypothesis supported by available data. Table 14.1 shows that the predictions of the financial data by the simple Wiener process, called as the Random Walk (RW) model, are as good as that of the sophisticated ARMA models.

The objective of traditional time series models is to define such parameters that minimize a deviation from the available data. One may call them as the best fit models. The goodness of fit is described by continuous parameters C called as state variables. For example, in the ARMA model (see expression (14.1)) the state variables are
$C = (a_i,\ i = 1, ..., p,\ b_j,\ j = 1, ..., q)$.

If the parameters remain constant, then models that fit best to the past data will predict the future data as well. Otherwise, the best fit to the past data can be irrelevant or even harmful for predictions. Therefore, one needs models which are not sensitive to the changes of parameters. Such models may predict the uncertain future better by eliminating the nuisance parts from the structure of the model.

Trying to solve this problem, one introduces a notion of the model structure. The model structure is determined by the Boolean parameters S called as structural variables. A structural variable is equal to unit, if the corresponding component of time series model is included. Otherwise, the structural variable is equal to zero.

For example, in the ARMA model $S = (s_i^a,\ i = 1, ..., p,\ s_j^b,\ j = 1, ..., q)$. Here $s_i^a = 1$, if the parameter a_i is included into the ARMA model. Otherwise, $s_I^a = 0$ [9]. We search for such structure S of the model that minimizes the prediction errors in the changing environment. To achieve this we divide available data $W = (w_t,\ t = 1, ..., T)$ into two parts $W_0 = (w_t,\ t = 1, ...T_0)$ and $W_1 = (w_t\ t = T_0 + 1, ..., T)$.

The first part W_0 is to estimate continuous parameters $C = C(S)$ that depends on Boolean structural parameters S. The estimates are obtained for a set of all feasible S by minimizing the least square deviation using data W_0.

The second part W_1 is used to select such S that minimize the least square deviation. This means that the second part W_1 is to estimate Boolean structural parameters.

Denote by $R_t(S, C, W)$ the predicted value of a model R with fixed parameters S, C using the data $(w_1, ..., w_{t-1}) \subset W$. The difference between the prediction and the actual data w_t is denoted by $\epsilon_t(S, C, W) = w_t - R_t(S, C, W)$. Denote by $C_0(S)$ the fitting parameters C which minimize the sum of squared deviations $\Delta_{0,0}(C, S)$ using the first data set W_0 at fixed structure parameters S.

$$\Delta_{0,0}(C, S) = \sum_{t=1}^{T_0} \epsilon_t^2(S, C, W), \tag{14.50}$$

$$C_0(S) = \arg\min_C \Delta_{0,0}(C, S). \tag{14.51}$$

We stabilize the structure S by minimizing the sum of squared deviations $\Delta_{1,0}(S)$ using the second data set W_1 and the fitting parameters $C_0(S)$ that were obtained from the first data set

$$\Delta_{1,0}(S) = \sum_{t=T_0+1}^{T} \epsilon_t^2(S, C_0(S), W), \tag{14.52}$$

$$S_1 = \arg\min_S \Delta_{1,0}(S). \tag{14.53}$$

[9]Note that describing a real data one needs diverse structures including a number of different models, not just one specific model as in this illustrative example.

This is a way to reach a tradeoff between the fitting parameters and the structural ones. The fitting parameters $C_0(S)$ provide the best fit to the first data set W_0 at fixed structure S. One stabilizes the structure S by minimizing the prediction errors for the second set of data W_1 using the fitting parameters $C_0(S)$. Here stabilization is achieved because the fitting parameters are defined by the first set of data W_0. The stabilized structure $S = S_1$ of R is obtained by eliminating unstable[10] parameters and parts of the time series.

We consider two data sets W_0 and W_1 just for simplicity. One may partition the data W into many data subsets $w_t \in W_k,\ t \in T_k,\ k = 1, ..., K,\ \cup_k W_k = W,\ \cup_k T_k = T$. In this case we minimize the sum

$$S_K = \arg\min_S \Delta(S), \tag{14.54}$$

where

$$\Delta(S) = \sum_{k,l} \Delta_{k,l}(S). \tag{14.55}$$

Here

$$\Delta_{k,l}(S) = \sum_{t \in T_k} \epsilon_t^2(S, C_l(S), W). \tag{14.56}$$

Note, that dividing the data W into many parts one may obtain sequences T_k too short for the meaningful estimate of parameters C_k, if T is not large. If T is large, one may expect that most of the fitting parameters C_k would be different in different data subsets W_k. Thus, they will be eliminated by the stabilization procedure (14.54). Therefore, $K = 2$ seems a reasonable number, at the first stabilization attempt. One may try $K > 2$ later.

The idea of the structural stabilization follows from the following observation. The best estimate of time series parameters, using a part W_0 of the data $W = W_0 \cup W_1$, is optimal for another part W_1 only if all the parameters remain the same. Otherwise, one may obtain a better estimate by elimination of the changing parameters from the model. For example, in the case of changing parameters ($a_i,\ i = 2, ..., p,\ b_j,\ j = 1, ..., q$) of the ARMA model, the best prediction is obtained by elimination of all these parameters, except $a_1 = 1$ (see Table 14.1).

---

[10]The parameters are considered as unstable if they change too much in different data sets.

10.2 SIMPLE EXAMPLE

Consider, for illustration, a simple example.

$$w_t = a_1(t)w_{t-1} + a_2(t)w_{t-2} + \epsilon_t, \ t = 3, 4, 5. \tag{14.57}$$

The observed values are $w_1 = -1$, $w_2 = w_3 = 1$, $w_4 = 2$, $w_5 = 1$. The first data set $W_0 = w_i, ..., w_4$ is to estimate continuous parameters $C = (a_1, a_2)$. The second data set $W_1 = w_5$ is to estimate Boolean parameters $S = (s_1^a, s_2^a)$, $s_i^a = 0, 1$. Equality to zero $s_i^a = 0$ suggests the elimination of the continuous parameter a_i. There are three feasible S, namely $S_1 = (0, 1)$, $S_2 = (1, 0)$, $S_3 = (1, 1)$[11].

Assume that unknown parameters depend on t this way

$$a_1(t) = \begin{cases} 2, & \text{if } t = 3, \\ 1, & \text{if } t = 4, \\ 0, & \text{if } t = 5, \end{cases} \tag{14.58}$$

$$a_2(t) = 1,$$
$$\epsilon(t) = 0.$$

In the case $S = S_3$, the least square estimates are $a_1(S_3) = 1.5$ and $a_2(S_3) = 0.5$. The prediction is $w_5(S_3) = 3.5$.
If $S = S_1$, the least squares estimate of the only remaining parameter is $a_2(S_1) = 0.5$. The prediction is $w_5(S_1) = 0.5$.
In the case $S = S_2$, the least square estimate is $a_1(S_2) = 1.5$ and the prediction is $w_5(S_2) = 3.0$.

The best prediction $w_5(S_1) = 0.5$ is provided by the structure $S_1 = (0, 1)$. The reason is obvious: this structure eliminates the highly unstable parameter a_1. Applying the structural stabilization one eliminates the nuisance parameter $a_1(t)$ and simplifies AR model (14.57)

$$w_t = a_2(t)w_{t-2} + \epsilon_t, \ t = 3, 4, 5. \tag{14.59}$$

Note, that we eliminate the larger parameter a_1 ($a_1(S_2) = 1.5 > a_2(S_2) = 0.5$) because it changes (see expression (14.58)).

[11]The elimination of both continuous parameters a_1 and a_2 is unfeasible.

10.3 EXAMPLES OF STRUCTURAL OPTIMIZATION WITH EXTERNAL FACTORS

Various examples of structural optimization are described in (Mockus, 1997). Here the structural optimization is considered in time series models with external factors.

If predictions depend on several factors, the multi-dimensional ARMA should be used. Denote by $\nu(Mt),\ t \geq p/M$ the main statistical component and by $\nu(Mt-i),\ i=1,..,p-1$ the external factors. One extends the traditional ARMA model this way

$$\nu(Mt) = \sum_{i=0}^{p-1} a_i \nu(Mt-i-1) + \sum_{j=0}^{q-1} b_j \epsilon(M(t-j-1)) + \epsilon(Mt), \quad (14.60)$$

$$p/M \leq t \leq T0$$

Here the number of the Auto Regressive (AR) components is denoted by p and the number of the Moving Average (MA) ones is denoted by q. The continuous variables a and b define the state of the ARMA model. We call them the state variables. The discrete parameters p, q, and $t0$ define the structure. One calls them the structural variables. The structural variable $t0$ defines the time when one starts scanning the time series for the optimization of state variables a, b. Denote by $T0$ the scanning end.

One minimizes the squared deviation at fixed structural variables $p, q, t0$

$$\Delta_{00}(p,q,t0) = \sum_{t=t0}^{T0} \epsilon(Mt)^2,\ t0 \geq p/M. \quad (14.61)$$

Here $T0 < T1 < T$ and $\epsilon(Mt)^2$ is from expression (14.60). Denote the optimal values $a = a(p,q,t0)$ and $b = b(p,q,t0)$.

If b is fixed, the optimal values of $a = a^b$ are defined by a system of linear equations. These equations follows from the condition that all the partial derivatives of sum (14.60) are equal to zero $\partial\Delta_{00}/\partial a_i = 0,\ i = 0,...,p-1$. Therefore, to obtain the least squares estimates $a_i = a_i^b,\ i=0,...,p-1$ at given b one solves p linear equations with p variables $a_i,\ i=0,...,p-1$ (see Chapter 3. for details).

To optimize discrete structural variables $p, q, t0$ one uses another data set. It starts at $T0$ and ends at $T1$. During optimization of structural variables one keeps the best fitting values of state variables $a = a(p, b, t0)$ and $b = b(p, q, t0)$. These values are obtained by minimization of the sum $\Delta_{01}(p, q, t0)$. The data is from $t0$ to $T0$. Here the sum

$$\Delta_{01}(p,q,t0) = \sum_{t=T0}^{T1} \epsilon(Mt)^2. \tag{14.62}$$

11. EXAMPLES OF SQUARED RESIDUALS MINIMIZATION

11.1 MULTI-MODALITY EXAMPLES

Consider these examples:

- exchange rates of \$/£, DM/\$, yen/\$, and franc/\$,
- closing rates of stocks of AT&T, Intel Corporation, and Hermis bank[12],
- London stock exchange index,
- daily call rates of a call center.

The following figures show the data and optimization results. The optimization results show how least square deviations depend on parameters b of ARMA models. The $3D$ optimization results are in two forms: as surfaces and as contours.

Figures 14.4, 14.5, and 14.6 consider exchange rates of \$/£ and DM/\$. Figures 14.7, 14.8, and 14.9 consider exchange rates of yen/\$ and franc/\$. Figures 14.10 reflect closing rates of AT&T (top) and Intel Co.(bottom) stocks.
Figures 14.13 , 14.14, and 14.15 consider the London stock exchange index.
Figures 14.16, 14.17, and 14.17 shows stock rates of the Hermis bank and optimization results.
Figures 14.18 and 14.19 show the daily call rates of a call center and illustrate optimization results.

[12] A small Lithuanian bank.

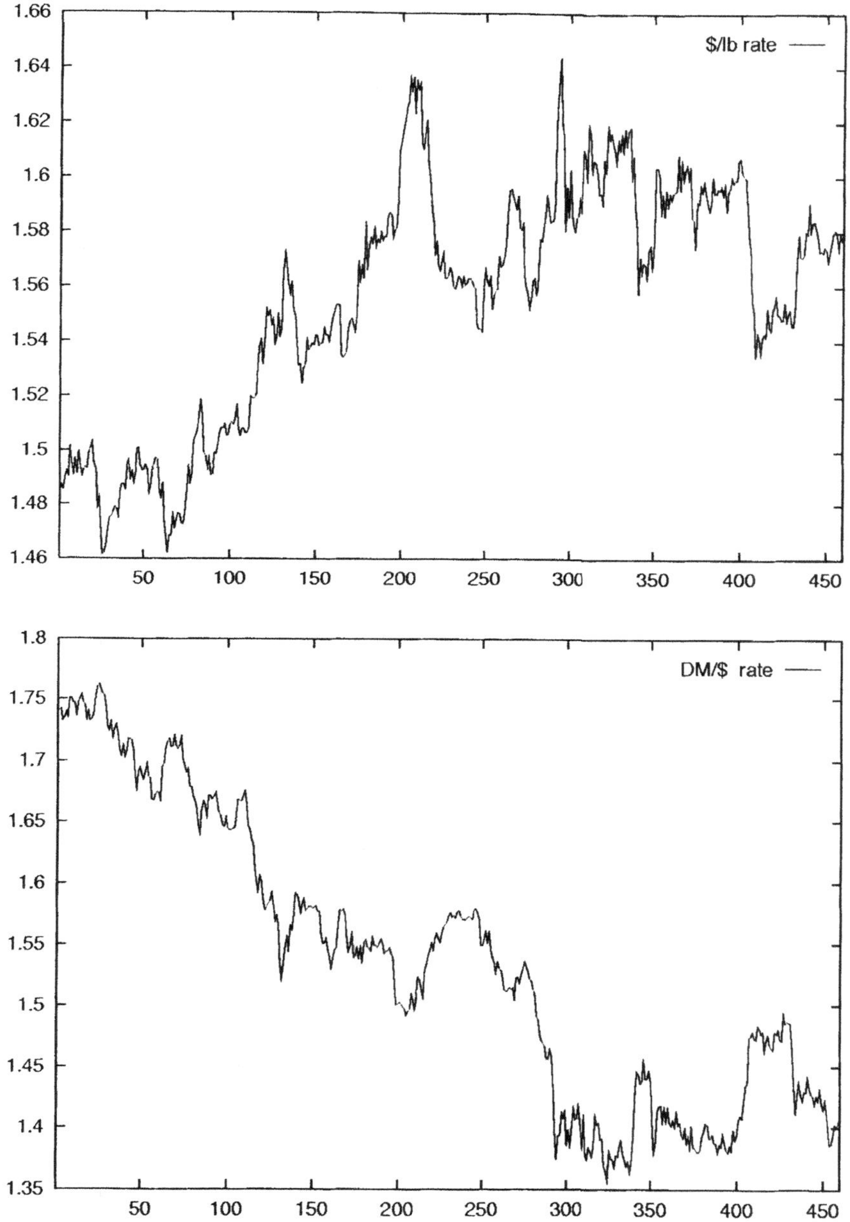

Figure 14.4. Daily exchange rate of $/£(top figure) and DM/$ (bottom figure) starting from September 13, 1993.

To estimate unknown ARMA parameters we minimize a log-sum of squared residuals defined by expression (14.45). Estimates of parameters

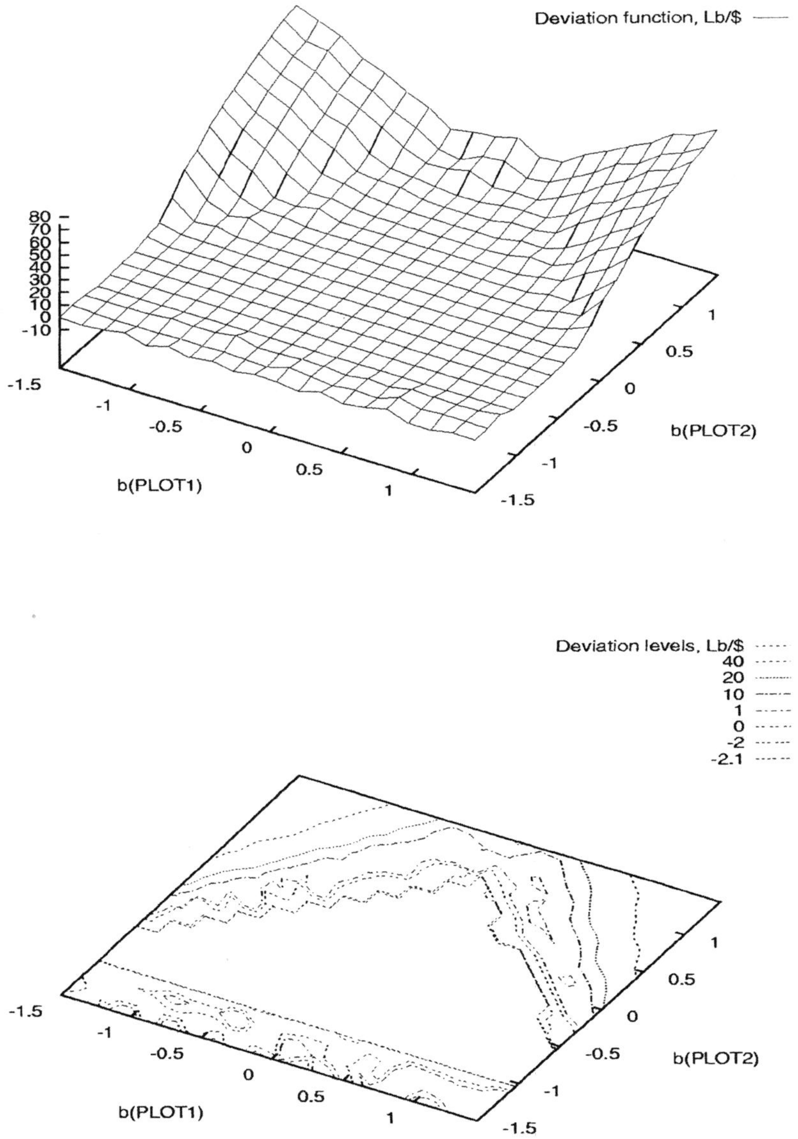

Figure 14.5. Exchange rates of £/$: surface f depending on parameters b_0, b_1 (top), contours of f depending on parameters b_0, b_1 (bottom).

a are from expression (14.13). In most of the cases, figures show the multi-modality of log-sum (14.45), as a function of parameters b_0, b_1.

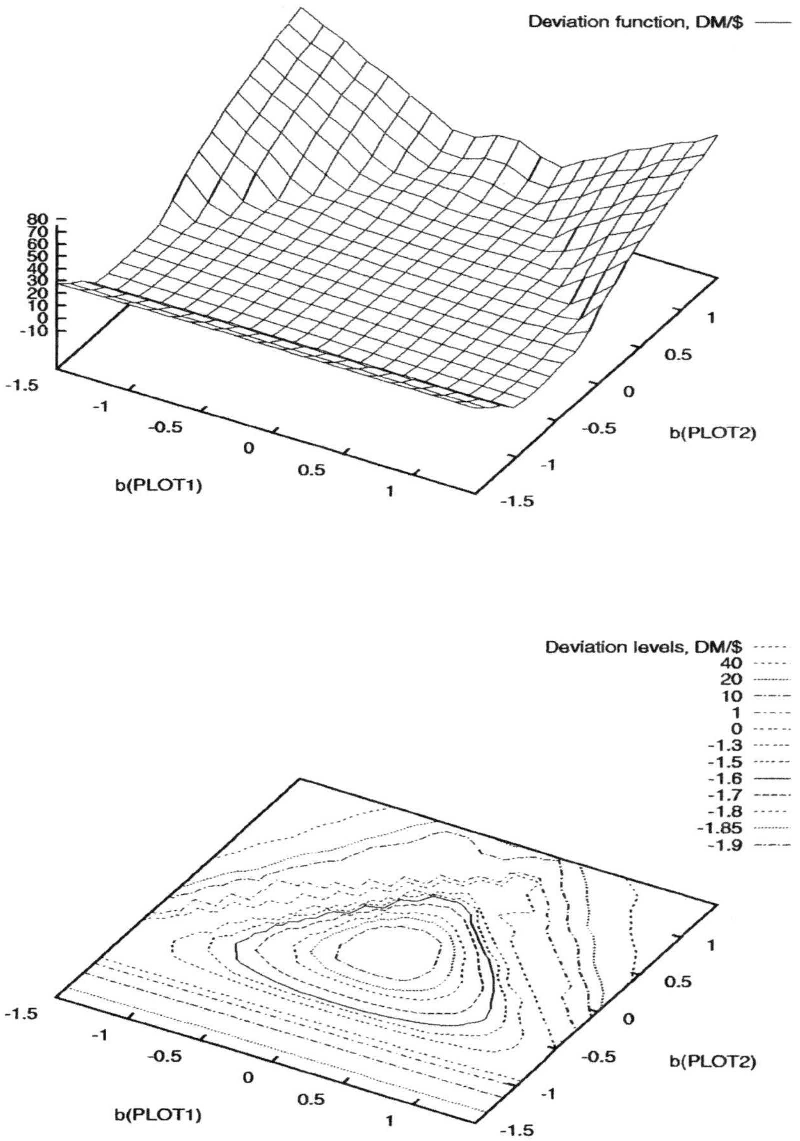

Figure 14.6. Exchange rates of DM/$: surface of f depending on parameters b_0, b_1 (top), contours of f depending on parameters b_0, b_1 (bottom).

Areas in vicinity of the global minima, often appear flat. A reason is that differences between values of the deviation function f in an area

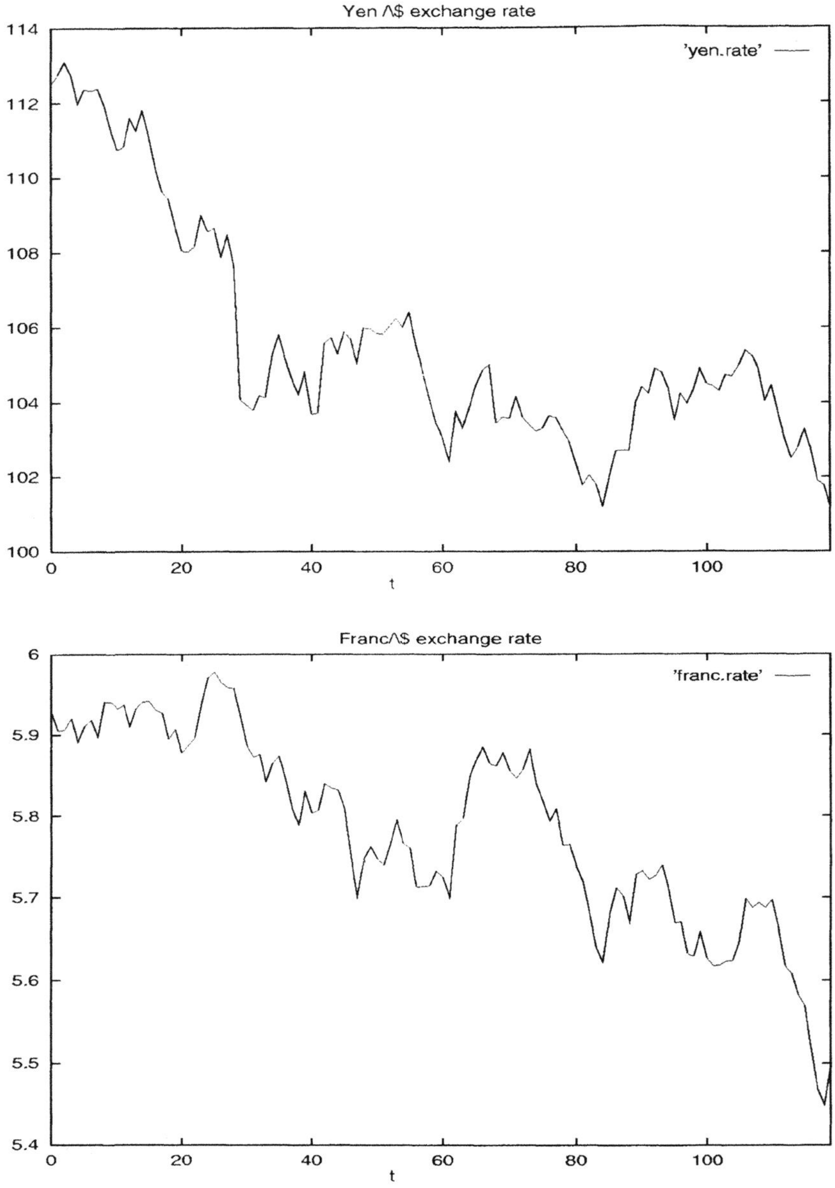

Figure 14.7. Exchange rates: yen/\$ (top), franc/\$ (bottom).

around the global minimum are smaller as compared with these outside this area (see Figures 14.15).

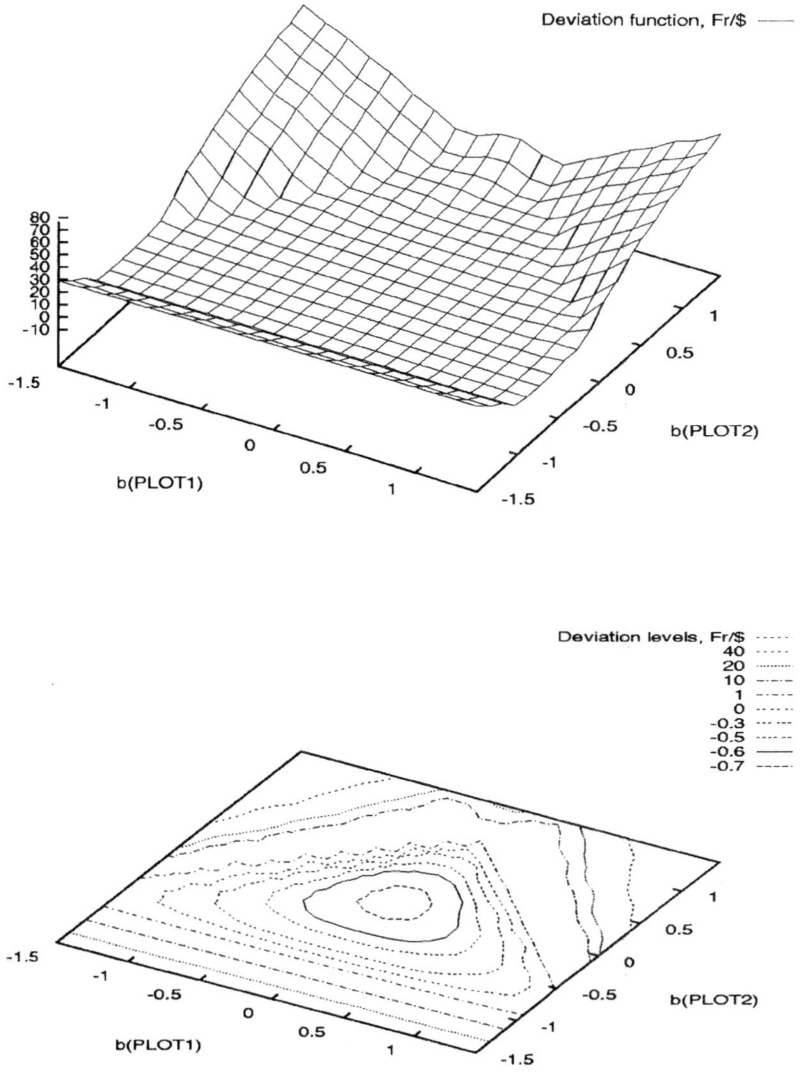

Figure 14.8. Exchange rates of fr/$, surface f depending on parameters b_0, b_1 (top), contours of f depending on parameters b_0, b_1 (bottom).

11.2 OPTIMIZATION RESULTS

The ARMA model optimization results are the points b and a (see Figure 14.1). These results were defined using a sequence of two global

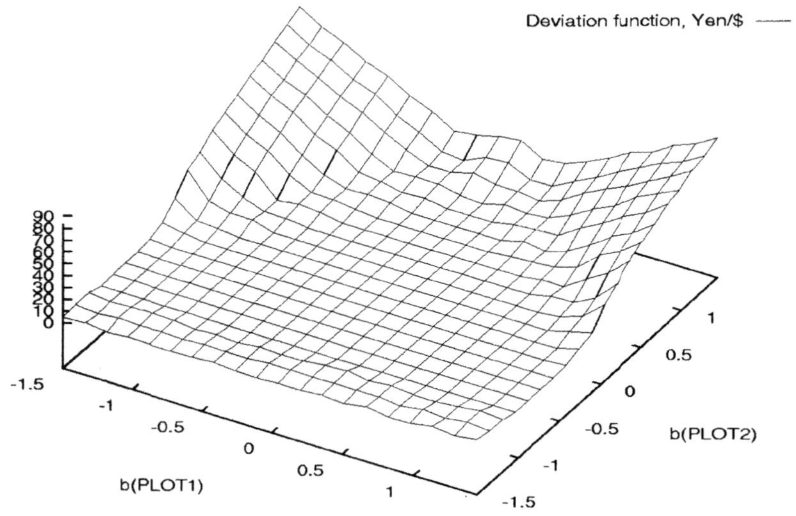

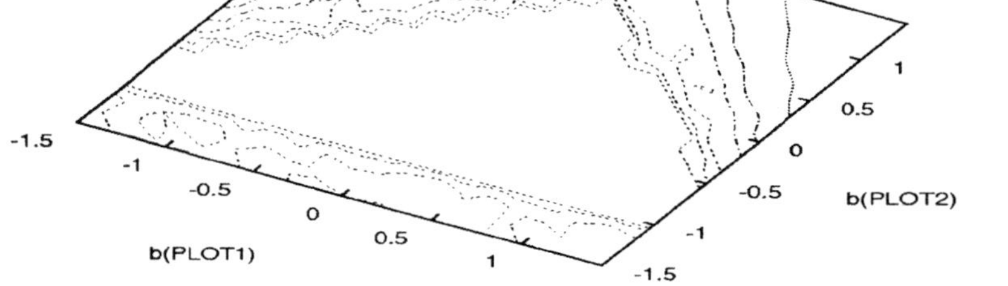

Figure 14.9. Exchange rates of yen/$, surface f depending on parameters b_0, b_1 (top), contours of f depending on parameters b_0, b_1 (bottom).

methods called $BAYES1$ and $EXKOR$ (Mockus et al., 1997). $BAYES1$ denotes a search by a multi-dimensional Bayesian model (Mockus et al., 1997). The best result obtained by $BAYES1$ after fifty iterations is a starting point for local optimization. The local optimization is by one-

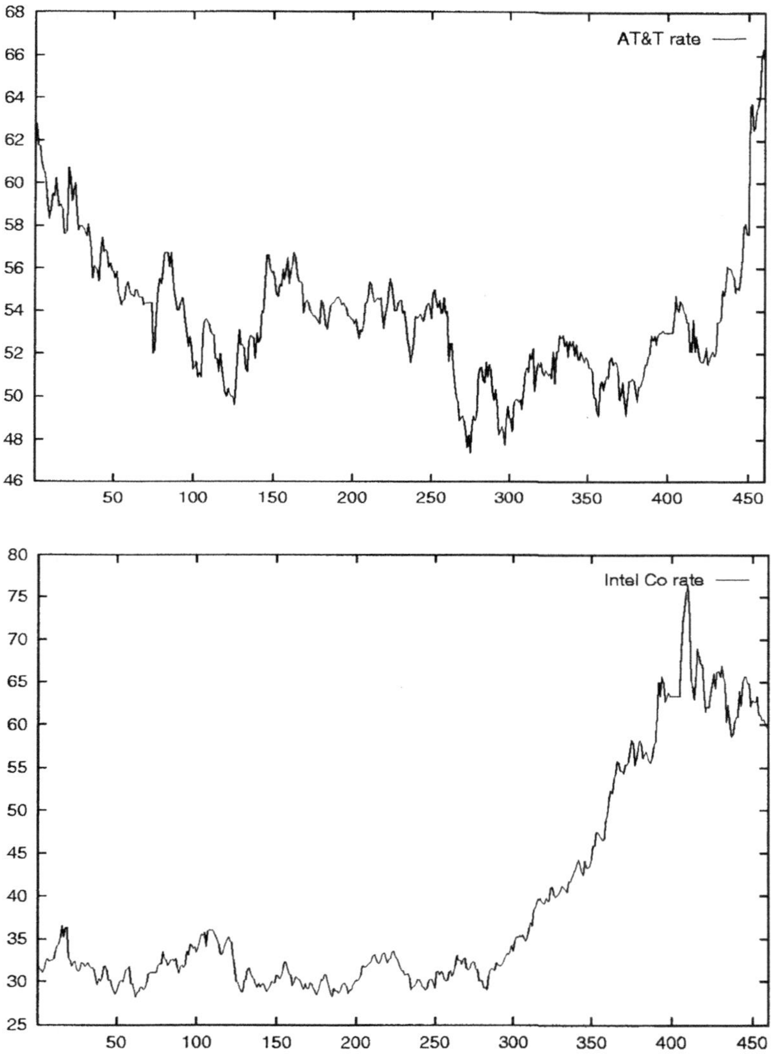

Figure 14.10. AT&T (top) and Intel Co.(bottom) stocks closing rates starting from August 30, 1993.

dimensional coordinate search $EXKOR$ (Mockus et al., 1997), using sixty iterations.

The maximal number of Auto Regressive (AR) parameters p was $10 * M$. Here $M = 1$, if no external factors are involved. The optimal number p is defined by structural stabilization (see Chapter 10.). Only two Moving Average (MA) parameters $q = 2$ were considered while

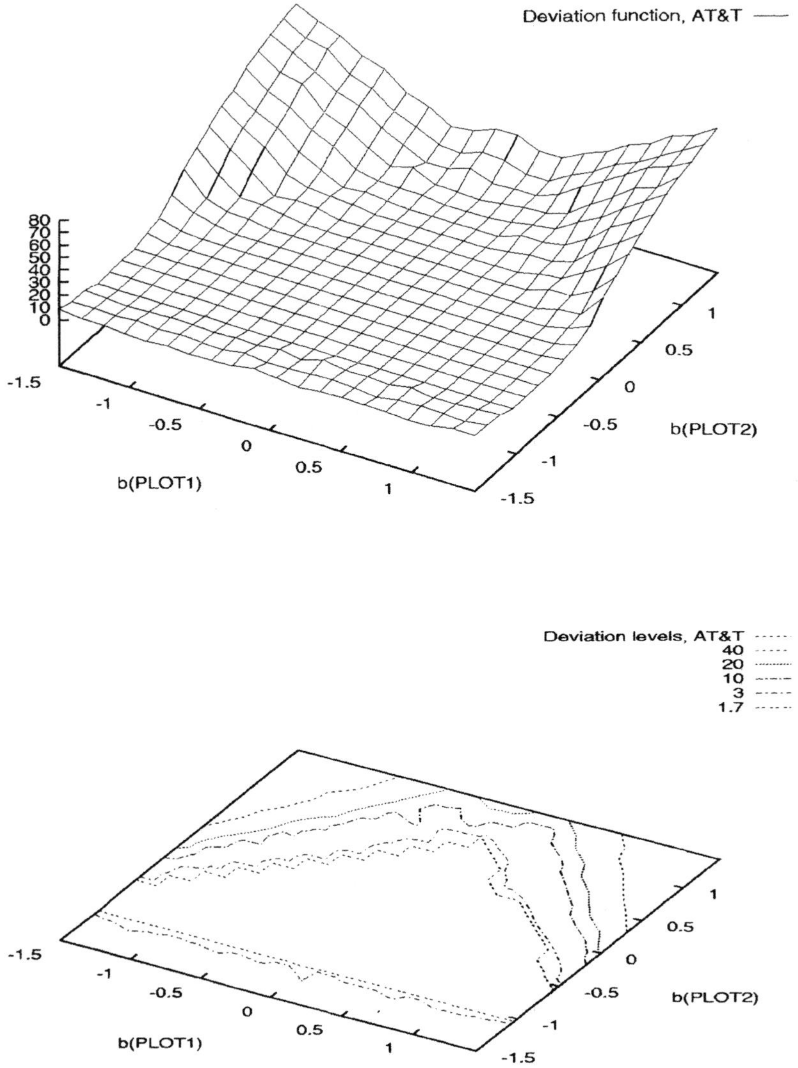

Figure 14.11. Closing rates of AT&T stocks: surface f depending on parameters b_0, b_1 (top), contours of f depending on parameters b_0, b_1 (bottom).

plotting surfaces and contours. The results of Table 14.1 were obtained by optimization of both structural variables p and q.

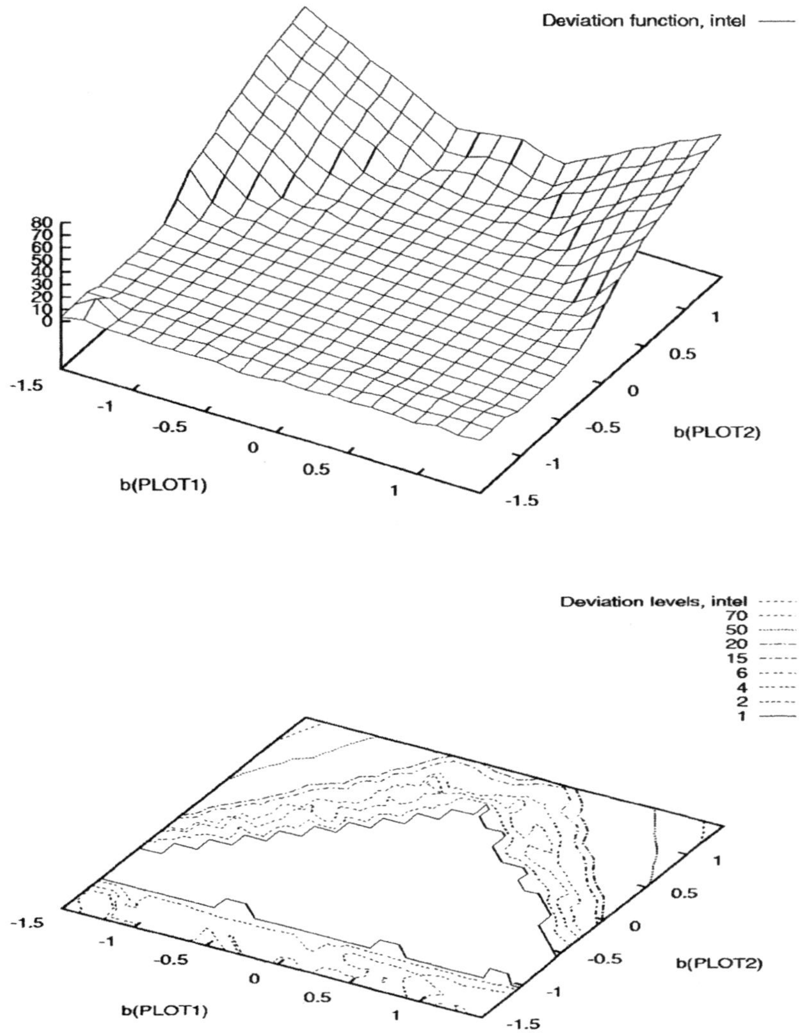

Figure 14.12. Closing rates of Intel Co stocks: surface of f depending on parameters b_0, b_1 (top) contours of f depending on parameters b_0, b_1 (bottom).

The objective of this part of research is to show a multi-modality arising in the prediction problem. Therefore, to save the computing time, the global optimization is carried out approximately, using not many iterations. The results of global optimization are starting points for local optimization. The reason is that the squared deviation as a

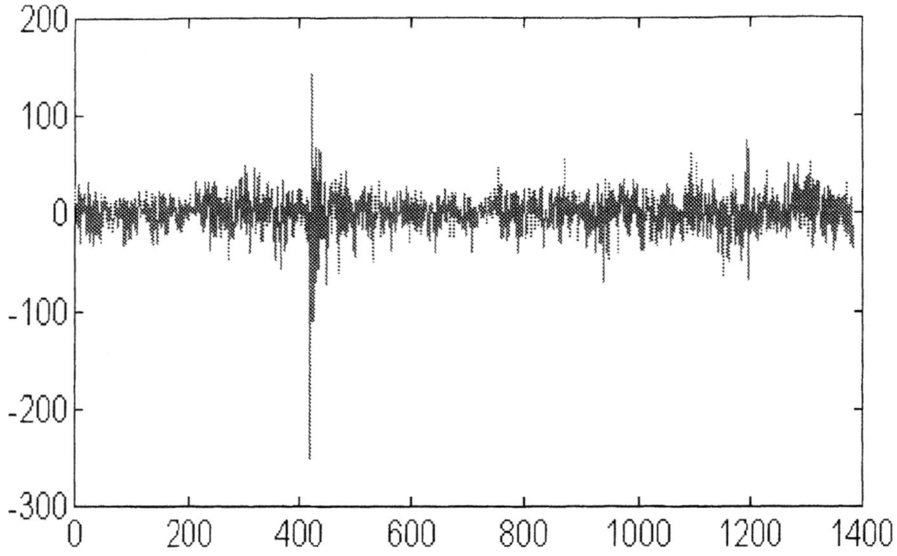

Figure 14.13. London stock exchange index.

function of parameters b becomes uni-modal near the global minimum (see Figures 14.15). Therefore, the results are at least as good as those obtained by the traditional local optimization.

The high-accuracy global optimization is very expensive. In the global optimization, the computing time is an exponential function of accuracy m, in the sense that $\epsilon \leq 2^{-m}$. Terefore, the problem of future investigations is how to balance computing expenses and accuracy of estimates. This task is important in both the time series prediction and the global optimization.

The investigation of multi-modality of squared deviation and variability of the parameters is the natural first step. The multi- modality is involved in non-linear regression models, including the ARFIMA ones[13].

[13]Meaning that the sum (14.45) of the ARFIMA model is a non-linear function of the parameters b and d.

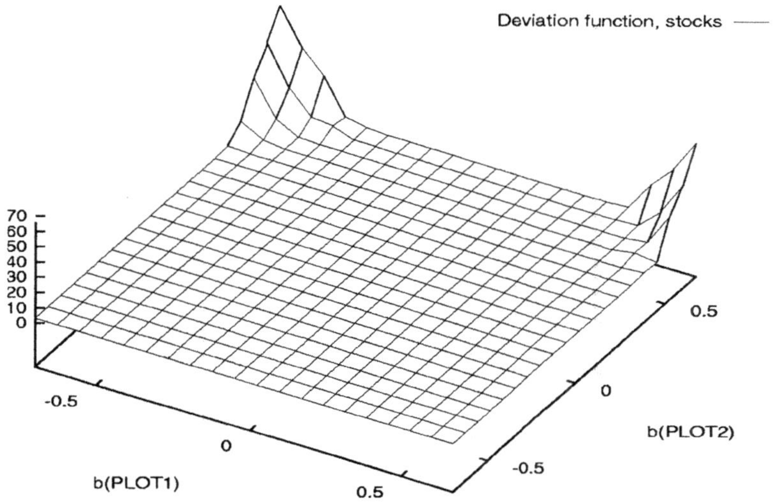

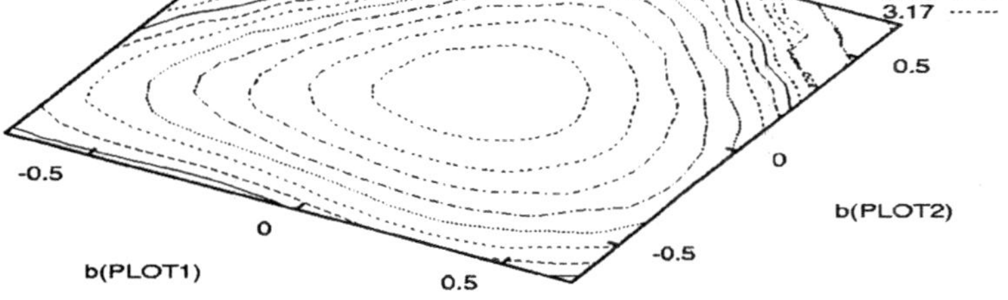

Figure 14.14. London stock exchange index, surface of f depending on parameters b_0, b_1 (top), contours of f depending on parameters b_0, b_1 (bottom).

12. SOFTWARE EXAMPLES

12.1 C VERSION OF ARMA SOFTWARE (ARMAC)

The call rates depend on several factors what is usual in statistical prediction problems. That is the main difference from the traditional

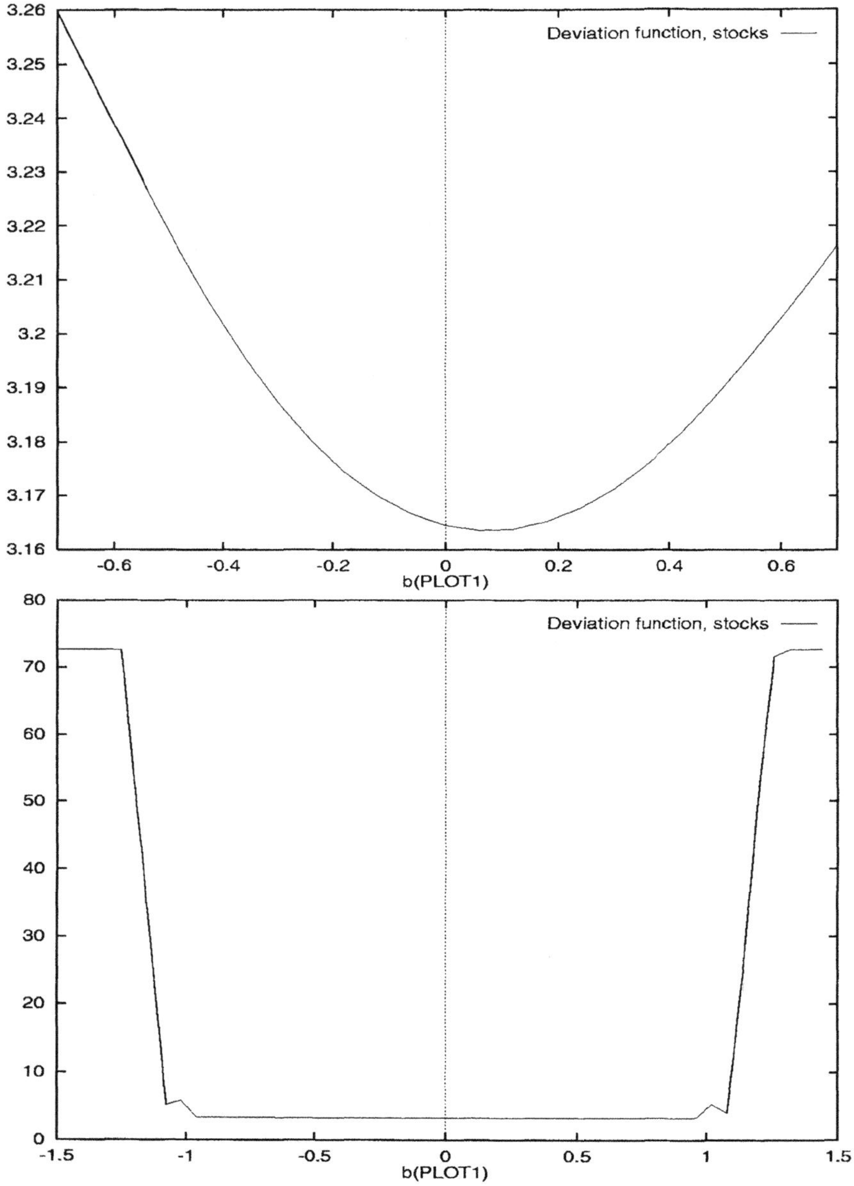

Figure 14.15. London stock exchange index: deviation f depending on parameter b_0, uni-modal part (top), multi-modal part (bottom).

ARMA and ARFIMA software for the exchange rate prediction (Mockus et al., 1997). The ARFIMA software and applications are described in (Mockus et al., 1997).

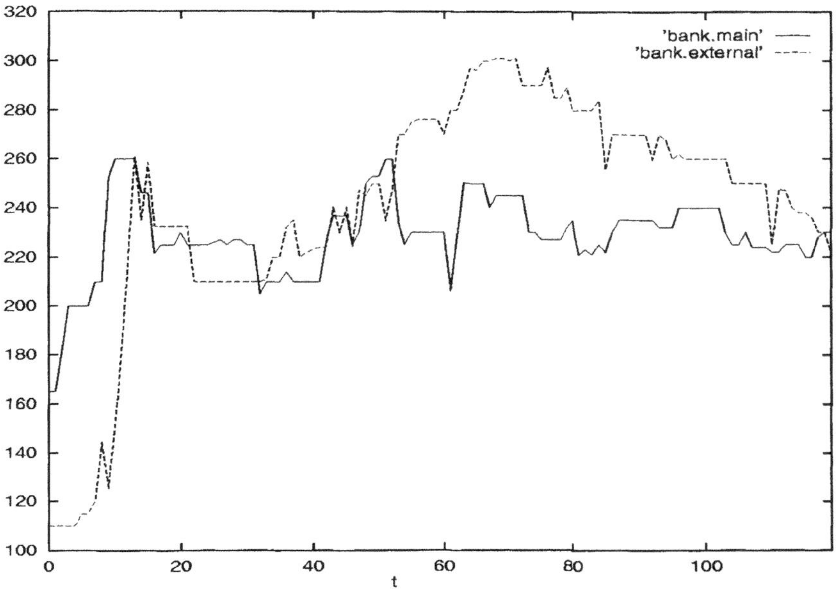

Figure 14.16. Closing rates of stocks of the Hermis bank (private) and the Lithuanian Savings bank (state owned).

We consider a version of an extended ARMA model that includes external factors (see expressions 14.28 and 14.29). The programs are in the files 'main.C', 'fi.C', and 'fitimeb.h' on the web-site (see Section 4.). The first version of ARMA software is designed for data sets with no future data. That means that no future factors are not known. Therefore, the external factors are treated as missing data. It is assumed that future values of external factors are equal to the last ones. In the ARMA software under development, this is considered as the default case. In the new software version, the known future values of external factors will replace the default ones.

The results of a simple test are in the files 'test.out' and 'test.progn.out'. The data is in the file 'test.data', the initiation file is 'test.ini'. The results of the call rate example are in the files 'call.out' and 'call.progn.out'. The data is in the file 'call.data'. The initiation file is 'call2.ini'. The names of the data files are referred as INP in the initiation file INIFILE. The software is compiled and run this way:

```
Compile by 'make onestep',
Run by 'onestep > results.out'
```

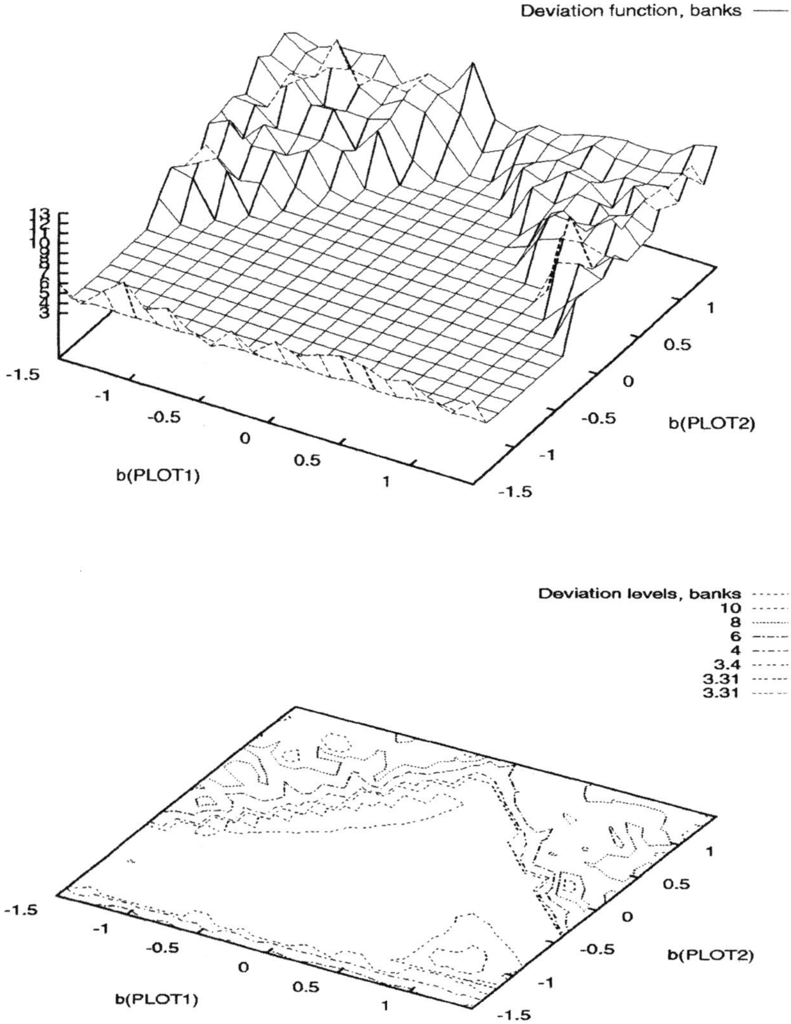

Figure 14.17. Closing rates of Hermis bank stocks: surface of f depending on parameters b_0, b_1 (top) contours of f depending on parameters b_0, b_1 (bottom).

ILLUSTRATIVE EXAMPLE OF ARMA SOFTWARE: STOCK RATE PREDICTION

Figures 14.20 and 14.21 show the part of the file 'fitimeb.h' defining the control parameters and data files.

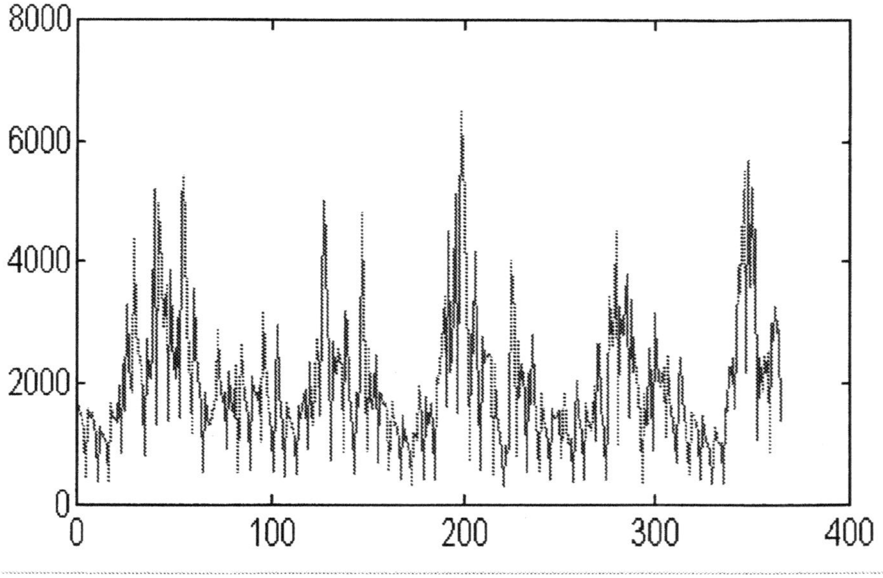

Figure 14.18. **Call rates.**

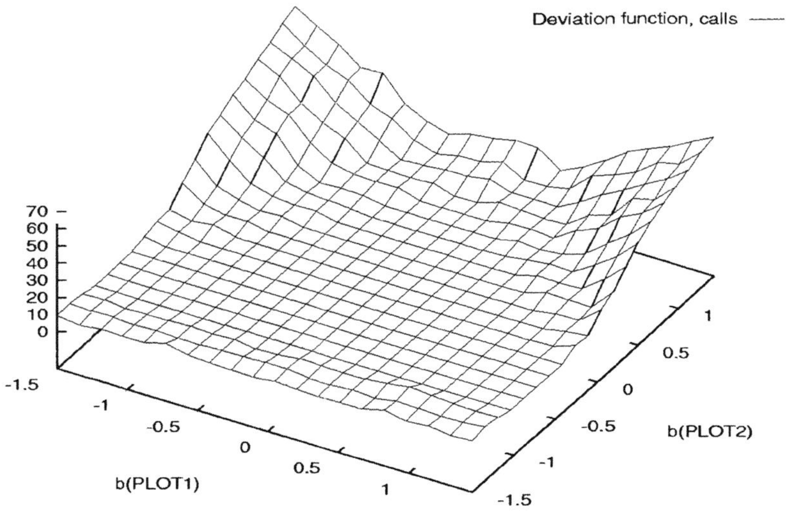

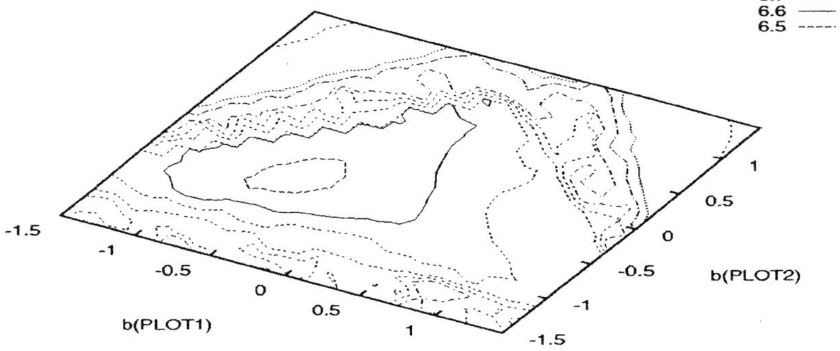

Figure 14.19. Call rates: surface of f depending on parameters b_0, b_1 (top), contours of f depending on parameters b_0, b_1 (bottom).

```
#define RAND() drand48()
#define S 0 /* number of rows of matrix c */
#if S == 0
#define S1 1
#else
#define S1 S
#endif
#define R 0 /* number of columns of matrix c */
#define K 5 /*number of "multi-step" repetitions*/
#define W 0 /*W 0 means the one-step
 structural optimization,
 W 1 defines the multi-step one*/
#define V 1 /*V 1 means with the multi-step prognoses,
V 0 means without*/
#define F  1 /*indicator of variance,
 F 1 involves variance*/
#define EPS 0/*indicator of residual printing*/
#define INP  0 /*indicator of input control*/
#define SA  0 /*indicator of simulated annealing */
#define PL  0/*plotting dimension*/
#define PLOT1 0 /*first plotting coordinate is b[PLOT1}*/
#define PLOT2 1/*second plotting coordinate is b[PLOT2}*/
#define A1 -1.5 /*lower bound of  b[PLOT1}*/
#define B1 1.5/*upper bound of  b[PLOT2}*/
#define A2 -1.5 /*lower bound of  b[PLOT2}*/
#define B2 1.5/*upper bound of  b[PLOT2}*/
#define DN 50 /*number of plotting steps*/
#define ST 10000. /*temperature of simulated annealing */
#define SI 100 /*number of
simulated annealing iterations*/
#define Ps M /*starting number of AR parameters*/
#define Qs 0 /*starting number of MA parameters*/
#define Pmin M /*minimal number of AR parameters*/
#define Qmin 0 /*minimal number of MA parameters*/
#define Pmax 2*M  /*maximal number of AR parameters*/
#define Qmax 2 /*maximal number of MA parameters*/
```

Figure 14.20. Example of the file 'fitimeb.h', part 1.

```
#define T 120*M /* number of data entries in DATAFILE
(divisible by M)*/
#define TO T/3 /*TO<T number of  entries
for a and b optimization (divisible by M)*/
#define T1 2*TO /*T1>=TO, number of  entries
for P and Q optimization(divisible by M)*/
#define TR -1*M /*TR <= T, TR/M is the  number of
the first line in DATAFILE
used for simple regression
(negative TR prints no regression)*/
#define TE 25*M/*TE>TR, TE/M is the number of
the last line for regression*/
#define INIFILE "bank2.ini"
#define M 2 /*number of factors*/
#define MAXCOLS  120
#define DM 200 /*array size*/
#define MAX_B_BOUND 1.0
/* -MAX_B_BOUND <= b[i] <= MAX_B_BOUND */
#define MAX_C_BOUND 1.0 /* -MAX_B_BOUND <= c[i][j]
<= MAX_B_BOUND */
//#define LOCAL_METH NLP
#define LOCAL_METH EXKOR
//#define GLOBAL_METH EXKOR
//#define GLOBAL_METH GLOPT
#define GLOBAL_METH BAYES1
//#define LOCAL_METH GLOPT
#define GLOPT_MAX_IT 800  /* glopt IT */
#define GLOPT_LT 100 /* glopt LT */
#define GLOPT_MAXL 200 /* glopt MAXL */

#define NLP_MAX_IT 10 /* nlp IT */
#define NLP_M 0 /* nlp M  */
#define NLP_ME 0  /* nlp ME */

#define BAYES1_MAX_IT   5*M /* bayes1 IT */
#define BAYES1_LT  5 /* bayes1 LT */

#define EXKOR_MAX_IT  6*M  /* exkor IT */
#define EXKOR_INIT_POINTS 6 /* exkor LT */
```

Figure 14.21. Example of the file 'fitimeb.h', part 2.

The control parameters and the optimization results (in the case INP 0) are shown in Figure 14.22.

```
Number of Factors M 2
Multi-step prognoses indicator V 1
Number of prognoses repetitions K 5
indicator of input control INP 0
the first line in datafile for regression, TR  -2
the  last line in datafile for regression, TE  50
starting number of AR parameters Ps 2
minimal number of AR parameters Pmin 2
minimal number of MA parameters Qmin 0
maximal number of AR parameters Pmax 4
maximal number of MA parameters Qmax 2
number of data entries T 240
number of  entries for a and b optimization T0 80
number of  entries for P and Q optimization T1 160
indicator of variance F 1
Global Max Iterat 10
Local Max Iterat 12
b-Bounds 1.000000e+00
Diskret Params Init
(Ps-AR params, Qs-MA params,ts-starting time) :
Ps Qs ts  = 2 0 0
Diskret  Params Step :  dP dQ dt = 2 1 1
Diskret  Params Max :  Pmax Qmax tmax = 4 2 0
Start AR:  l=0_Ps a[l] =  -9.152631e-02
Start AR:  l=0_Ps a[l] =  1.083756e+00
Start Results: opt delta_perc delta_mean=
1.140519e+01  4.830240e+01  6.171501e+00
Start Discret: Po Qo to   opt   delta_perc  delta_mean =
2 0 0 1.140519e+01 4.830240e+01 6.171501e+00
Start  AR:  l=0_Po  a[l] =  -9.152631e-02
Start  AR:  l=0_Po  a[l] =  1.083756e+00
BAYES,
EXKOR,
Finish Discret: Po Qo toP Q t0 opt delta_perc delta_mean=
2  0  0      2  0  0
1.140519e+01  4.830240e+01  6.171501e+00
Finish  ARo:  l=0_P  a[l] = -9.152631e-02
Finish  ARo:  l=0_P  a[l] = 1.083756e+00
Finish: sigma 2.054141e+01
Finish progn_err 2.438730e+01
```

Figure 14.22. Example of the file 'results.out.'

The comments in Figures 14.20 and 14.20 are short. Here are some additional explanations

- `#define RAND() drand48(),`

  defines the random number generator

- `#define S 0 /* number of rows of matrix c */`
  `#define S 0 /* number of rows of matrix c */`

  defines the bilinear component (see expression 14.35)

- `#define K 5 /*number of "multi-step" repetitions*/,`

  K 5 means that the multi-step predictions is repeated 5 times (see Section 9. and Figure 14.23)

- `#define W 0 /*W 0 means the one-step`
  `structural optimization,`
  `W 1 defines the multi-step one*/,`

  one-step structural optimization minimizes the average error of the "next day" predictions (see Section 10.3), multi-step one minimizes the average error of predictions for longer periods of time

- `#define V 1 /*V 1 means with the multi-step prognoses,`
  `V 0 means without*/`

  the zero value of the indicator V switches off the multi-step prediction (see Section 9.) and switches on the next day prediction (see Section 3.3 and Figure 14.23)

- `#define F  1 /*indicator of variance,`
  `F 1 involves variance*/,`

  the unit value of the indicator F means that the variance of the errors ϵ is estimated and included into the multi-step prediction process

- `#define EPS 0/*indicator of residual printing*/,`

  not used in this version

- `#define INP  0 /*indicator of input display*/,`

  if INP 1 then the input values of the predicted factor are printed with their numbers (see Figure 14.24)

- `#define SA  0 /*indicator of simulated annealing */`

  if SA 1 then optimization of the parameters p and q are performed using the simulated annealing method, otherwise the exhaustive search is used

```
progn t  ymin[M*t]  yav[M*t]  ymax[M*t]
 80    2.212638e+02    2.415062e+02    2.549196e+02

progn t  ymin[M*t]  yav[M*t]  ymax[M*t]
81    2.479502e+02    2.657038e+02    2.954255e+02

progn t  ymin[M*t]  yav[M*t]  ymax[M*t]
82    2.327805e+02    2.725516e+02    3.380554e+02

progn t  ymin[M*t]  yav[M*t]  ymax[M*t]
83    1.911297e+02    2.686203e+02    3.272436e+02

progn t  ymin[M*t]  yav[M*t]  ymax[M*t]
84    2.059219e+02    2.715981e+02    3.346959e+02

progn t  ymin[M*t]  yav[M*t]  ymax[M*t]
85    1.989816e+02    2.731343e+02    3.315979e+02

progn t  ymin[M*t]  yav[M*t]  ymax[M*t]
86    1.998928e+02    2.742784e+02    3.425031e+02

progn t  ymin[M*t]  yav[M*t]  ymax[M*t]
87    1.946436e+02    2.708346e+02    3.392612e+02

progn t  ymin[M*t]  yav[M*t]  ymax[M*t]
88    1.870341e+02    2.722931e+02    3.534729e+02

progn t  ymin[M*t]  yav[M*t]  ymax[M*t]
89    1.647438e+02    2.618776e+02    3.345000e+02
```

Figure 14.23. A fragment of the file 'progn.out', ten day predictions.

- `#define PL  0/*plotting dimension*/,`

  if PL 1 then the values of objective function depending on the parameter b[PLOT1] will be written in the file 'plot.out',
  if PL 2 then the values of objective function depending on two parameters b[PLOT1] and b[PLOT2] will be written in the file 'plot.out' (see Figure 14.25) and plotted as Figures 14.17 using the "Gnuplot" system,
  if PL 0 there will be no file 'plot.out'

- `#define PLOT1 0 /*first plotting coordinate is b[PLOT1}*/`

```
0    165.00
 1     110.00
 2     165.00
 3     110.00
 4     181.00
 5     110.00
 6     200.00
 7     110.00
 8     200.00
 9     110.00
 10     200.00
```

Figure 14.24. A fragment of input data the file 'bank.out' when INP is set to unit.

```
-1.000000e+00 0.000000e+00 3.323284e+00
-1.000000e+00 0.000000e+00 3.323284e+00
-1.000000e+00 0.000000e+00 3.323284e+00
-1.000000e+00 0.000000e+00 3.323284e+00
-1.000000e+00 0.000000e+00 3.323284e+00
-1.000000e+00 0.000000e+00 3.323284e+00
-1.000000e+00 0.000000e+00 3.323284e+00
-1.000000e+00 0.000000e+00 3.323284e+00
-1.000000e+00 0.000000e+00 3.323284e+00
-1.000000e+00 0.000000e+00 3.323284e+00
```

Figure 14.25. A fragment of the ten entries the 'bank.plot.out' file.

`#define PLOT2 1/*second plotting coordinate is b[PLOT2}*/`,

defines which components of vector-parameter b are considered

- ```
  #define A1 -1.5 /*lower bound of  b[PLOT1}*/
  #define B1 1.5/*upper bound of  b[PLOT2}*/
  #define A2 -1.5 /*lower bound of  b[PLOT2}*/
  #define B2 1.5/*upper bound of  b[PLOT2}*/
  #define DN 50 /*number of plotting steps*/
  ```

  defines the range and the density of the plotting points

- ```
  #define ST 10000. /*temperature of simulated annealing */
  #define SI 100 /*number
   of simulated annealing iterations*/
  ```

defines the parameters of simulated annealing method (if SA is applied)

- ```
  #define Ps M  /*starting number of AR parameters*/
  #define Qs 0  /*starting number of MA parameters*/
  #define Pmin M /*minimal number of AR parameters*/
  #define Qmin 0  /*minimal number of MA parameters*/
  #define Pmax 2*M  /*maximal number of AR parameters*/
  #define Qmax 2  /*maximal number of MA parameters*/
  ```

  defines the initial, the minimal and the maximal number of parameters p and q in the structural optimization

- ```
  #define T 120*M /* number of data entries in DATAFILE
  (divisible by M)*/,
  ```

  defines the total number of entries in DATAFILE

- ```
  #define T0 T/3 /*T0<T number of  entries
  for a and b optimization (divisible by M)*/
  #define T1 2*T0 /*T1>=T0, number of  entries
  for P and Q optimization(divisible by M)*/,
  ```

  divides the DATAFILE (see Figures 14.26 and 14.16) into three parts: the first part for a,b optimization, the second part for p,q optimization, and the third part for testing the results

  ```
  #define TR -1*M /*TR <= T, TR/M is the  number of
  the first line in DATAFILE used for simple regression
  (negative TR prints no regression)*/
  #define TE 25*M/*TE>TR, TE/M is the number of
  the last line for regression*/,
  ```

  is used only in the case when the ARMA model of time series prediction is reduced to the linear regression model of diagnosis.

- ```
  #define INIFILE "bank2.ini",
  ```

  defines the input control file INIFILE (see Figure 14.27)

- ```
  #define M 2 /*number of factors*/,
  ```

  defines the number of factors, including the predicted and the external ones

- ```
  #define LOCAL_METH EXKOR,
  ```

  means that the EXKOR method is used for local optimization

- ```
  #define GLOBAL_METH BAYES1
  ```
  ```

```
         5786.09    180.00    224.00        250.00
40.00      20.25     85.00     16.00
         5796.10    180.00    222.00        225.00
40.00      20.25     85.00     16.00
         5806.11    180.00    222.00        247.50
40.00      20.00     80.00     16.00
         5816.12    181.00    225.00        247.00
40.00      20.00     80.00     15.99
         5826.13    182.00    225.00        240.00
40.00      20.00     84.50     15.99
         5836.16    185.00    225.00        238.00
36.00      20.00     85.00     15.95
         5846.17    188.00    220.00        238.00
36.00      20.00     85.00     14.36
         5856.18    188.00    220.00        236.00
36.00      20.00     85.00     14.30
         5866.19    190.00    228.00        230.00
36.00      20.00     85.00     15.45
         5876.20    195.00    230.00        230.00
36.00      20.00     85.00     15.00
         5886.23    190.00    230.00        221.00
36.00      20.00     85.00     15.50
```

Figure 14.26. A fragment of the ten recent days of the DATAFILE 'bank.data' inckuding the date code and data of seven main Lithuanian commercial banks.

```
INP bank.data
COL 3
COL 4
```

Figure 14.27. Example of the INIFILE 'bank2.ini.'

means that the BAYES1 method is used for global optimization

- `#define BAYES1_MAX_IT   5*M /* bayes1 IT */`
  `#define BAYES1_LT  5 /* bayes1 LT */`

  means that 5*M iterations and 5 initial iterations of the BAYES1 method are used

- `#define EXKOR_MAX_IT  6*M  /* exkor IT */`
  `#define EXKOR_INIT_POINTS 6 /* exkor LT */`

means that 6*M iterations and 6 initial iterations of the EXKOR method are used

ILLUSTRATIVE EXAMPLE OF ARMA SOFTWARE: SIMPLE TEST FILE

The test file is defined by 'test.ini':

```
INP test2
COL 1
COL 2
```

Here INP defines the data file 'test2'

```
1.   1.
2.   1.
1.   1.
2.   1.
1.   1.
2.   1.
1.   1.
2.   1.
1.   1.
```

Here COL 1 means that the first column of the file 'test2' should be considered as the factor to be predicted. COL 2 indicates the second column as an external factor.

A fragment of the optimization results is shown in the Figure 14.28. Here 'delta_perc' denotes the error in percents of the "Random Walk" prediction. 'delta_mean' is the average error. The results of optimization are printed as Po, Qo, ao, bo. Predicted results are in the file 'progn.out.old' (see Figure 14.29). The minimal $ymin$, the average yav, and the maximal $ymax$ predicted values are equal. Therefore, the prediction variance $progn$ is zero.

```
#define V 1 /*V 1 means with the multi-step prognoses,
V 1 means without*/
#define F  0 /*indicator of variance,
 F 1 involves variance*/
#define Ps M /*starting number of AR parameters*/
#define Qs 0 /*starting number of MA parameters*/
#define Pmin M /*minimal number of AR parameters*/
#define Qmin 0 /*minimal number of MA parameters*/
#define Pmax 2*M  /*maximal number of AR parameters*/
#define Qmax 0 /*maximal number of MA parameters*/
#define T 16 /* number of data entries in DATAFILE
(divisible by M)*/
#define T0 6 /* number of  entries
 for a and b optimization*/
#define T1 12 /* number of  entries
 for P and Q optimization*/
#define INIFILE "test.ini"
#define M 2 /*number of factors*/
Ps Qs ts  = 2 0 0
Diskret  Params Step :  dP dQ dt = 2 1 1
Diskret  Params Max :  Pmax Qmax tmax = 20 0 0
Start AR:  l=0_Ps a[l] =  3.000000e+00
Start AR:  l=0_Ps a[l] =  -1.000000e+00
Start Results: opt delta_perc delta_mean= 0.000000e+00
-1.000000e+02 0.000000e+00
Start Discret: Po Qo to   opt   delta_perc  delta_mean =
2 0 0
0.000000e+00 -1.000000e+02 0.000000e+00
Start  AR:  l=0_Po  a[l] =  3.000000e+00
Start  AR:  l=0_Po  a[l] =  -1.000000e+00
Finish Discret: Po Qo toP Q t0 opt delta_perc delta_mean=
2  0  0
2  0  0     0.000000e+00  -1.000000e+02  0.000000e+00
Finish  ARo:  l=0_P  a[l] = 3.000000e+00
Finish  ARo:  l=0_P  a[l] = -1.000000e+00
Finish: sigma 0.000000e+00
Finish progn 0.000000e+00
Finish: sigma 0.000000e+00
```

Figure 14.28. A fragment of the control parameters and optimization results of the simplest illustration.

```
progn t  ymin[M*t]  yav[M*t]  ymax[M*t]       6
1.000000e+00     1.000000e+00     1.000000e+00
progn t  ymin[M*t]  yav[M*t]  ymax[M*t]       7
2.000000e+00     2.000000e+00     2.000000e+00
progn 0.000000e+00
```

Figure 14.29. Example of the prediction results of the simplest illustration.

12.2 JAVA VERSION OF ARMA SOFTWARE (ARMAJ)

The Java version of ARMA software (ARMAJ) (Kuzminaite, 1999) implements the same algorithms as the C++ version ARMAC. The ARMAJ is on the web-sites (see Section 4.) and can be run by remote users.

USERS GUIDE

The applet 'index.html' is started by a browser, for example, by Netscape 4.6, or by the appropriate appletviewer. One clicks the button 'Show' (see the top Figure 14.30) to open the main window 'ARMA Frame' (see the bottom Figure 14.30). There are four buttons: 'File,' 'Input,' 'Options,' and 'Output' which open corresponding windows.

File is for data input.
There are two fields: 'INI File' and 'Working directory or URL.'
There is the button 'Browse...' and two options:
'Local file' and 'Local URL.'

The option 'Local file' activates the 'Browse...' button to select some local file.
The option 'Local URL' closes this button. Then the contents of fields 'INI File' and 'Working directory or URL' determine the data file.
The file name, for example, 'arma.ini,' is in the field 'INI File.'
The file 'arma.ini' controls the data input from the test file 'arma.test.'
The directory is in the field 'Working directory or URL.'
If the 'Local URL' option is on, then the directory is, for example, this:

```
http:/optimum.mii.lt/~jonas/armajavaj
```

The default URL is the applet directory.

Input is to change the default values of parameters (see the top Figures 14.31). The list of parameters is the same as in the C file 'fitimeb.h.'

Options are to select options (see the top Figure 14.32):

Output options defines the output file,

None means no output,

System output means console output,

Frame defines output into separate windows,

File outputs data into local files[14]

Graphic options involves graphics.

Output is for data output (see the bottom Figure 14.32):
one starts computations by clicking the button 'Calculate',
the message ' ARMA Frame: calculated' inicates the end.

Some input and output operations are not permitted by the 'Security Manager.' Using the 'appletviewer,' it is possible to bypass the 'Security Manager.' For example, in the file */.hotjava/properties* one writtes:

```
acl.read=/home/jonas/public\_html/armajavaj/arma.ini
acl.write=/home/jonas/public\_html/armajavaj/arma.out
```

That permits to read from 'arma.ini' and to writte into 'arma.out.'

The file *arma_all.jar* is the 'jar' archive including all the 'class' files.
source.jar or *source.zip* are archives of 'java' files.
'index.html' is a starting applet.
'arma.ini' is the input control file.
'armatest' is the test data file.

[14]This operation is not permited by browsers, as usuall.

ILLUSTRATIONS

The top Figure 14.30 shows the applet sign with the 'Show' button that starts ARMAJ. The botom Figure 14.30 shows the initial window where the input file 'arma.ini' is defined. button that starts the ARMAJ.

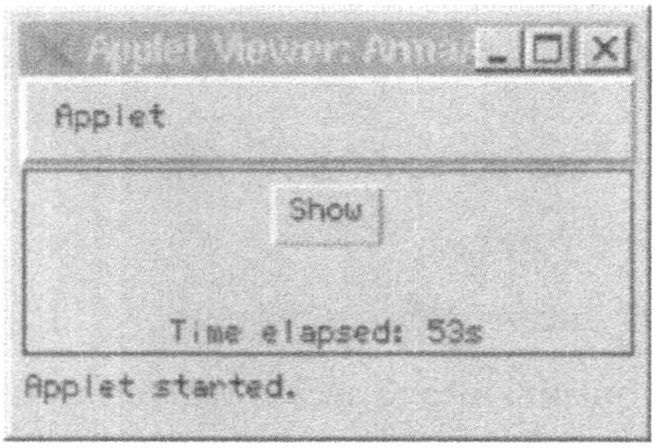

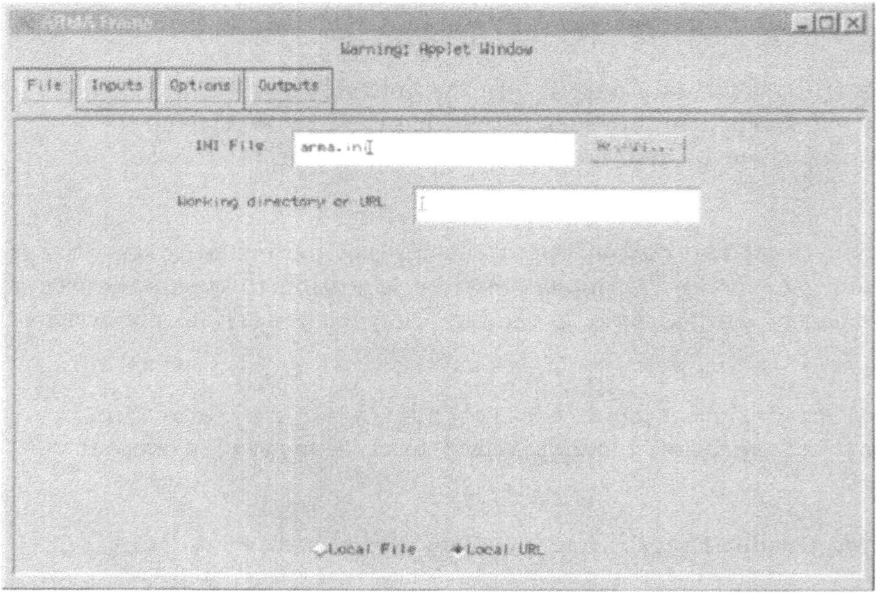

Figure 14.30. The applet sign (top figure), and the initial window defining the input file 'arma.ini'(bottom figure).

Figures 14.31 show the list of default values of control parameters, similar to those in the ARMA C++ version. The upper part of the list is on the top figure, the middle part is on the bottom one.

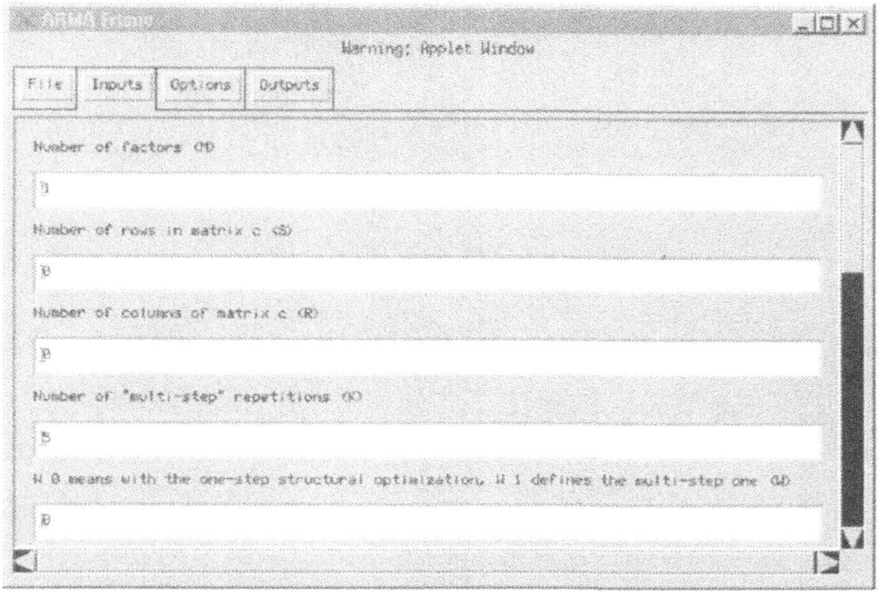

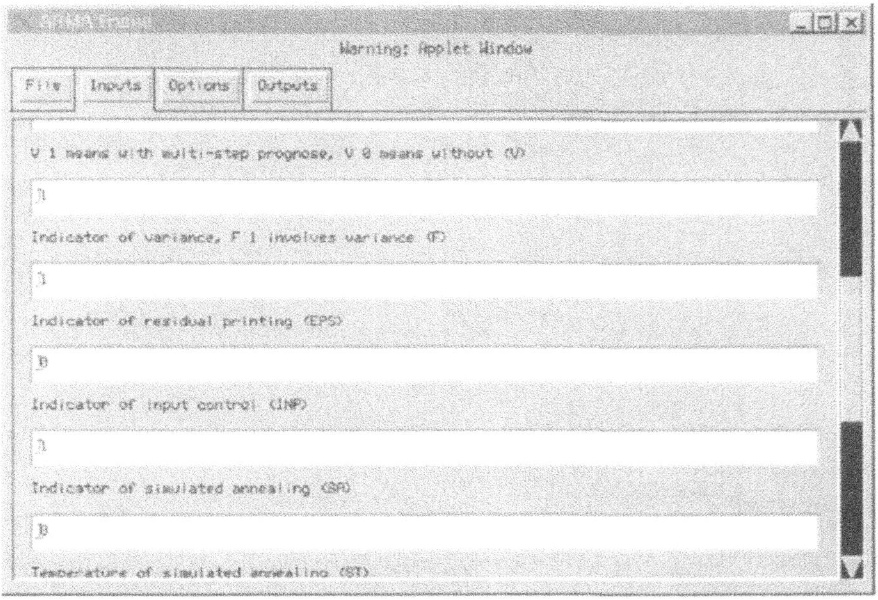

Figure 14.31. **The list of default values of control parameters, the upper part is on the top figure, the middle part is on the bottom one.**

The top Figure 14.32 shows the option window. The botom Figure 14.32 shows the last fragment of the output window. Here the results, similar to those in the ARMA C++ version, are written.

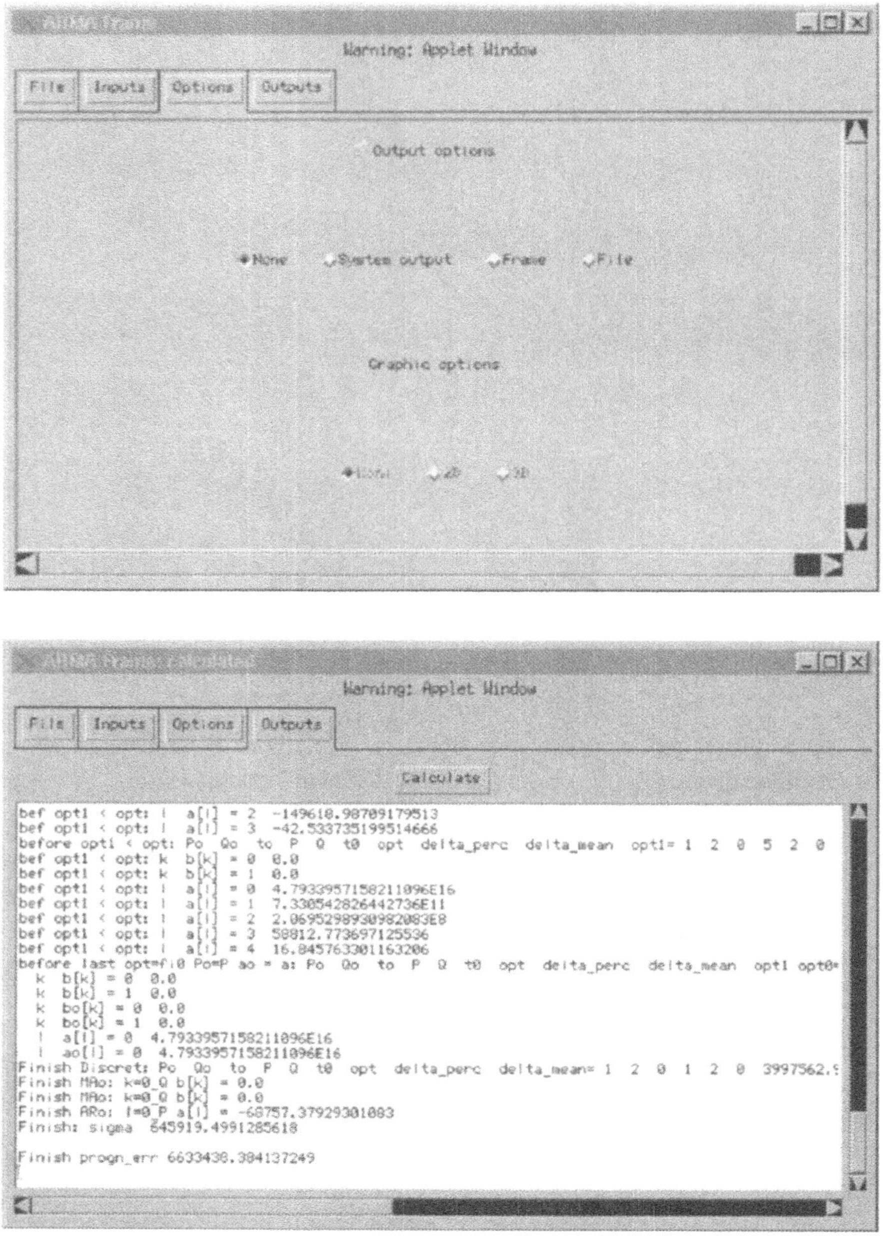

Figure 14.32. The option window (top figure) and the last fragment of the output window.

12.3 ANN SOFTWARE

A version of the ANN model (see expressions 14.31 and 14.34) is in the file *fi.C.anngauss* on the web-site (see Section 4.). *fi.C.anngauss* is designed as an objective function of the GMC global optimization system.

The software is compiled and run this way:

```
Extract files by 'tar -zxf gmc.tgz'
Rename the file 'fi.C.anngauss' by 'cp fi.C.anngauss fi.C'
Compile by 'make',
Run by './test'
```

Chapter 15

CALL CENTER MODEL

1. INTRODUCTION

1.1 OUTLINE

Call centers are important and rapidly developing commercial activities. Call centers serve customers by phone, by fax and by Internet. There are different call centers depending on their objectives and environments. However, investigating the call centers one encounters three problems:

- modeling;
  queuing systems are common tools while modeling call centers,

- optimization;
  one needs methods of stochastic and global optimization are to optimize these systems,

- prediction; parameters, which are needed for modeling and optimization, depends on predictions, predictions involve both the observed data and the expert knowledge.

1.2 ASSUMPTIONS, NOTATIONS, AND OBJECTIVES

One models call centers as queuing systems. The modeling system includes features for call rate predictions and service optimization. One assumes that

- incoming calls are united into one stream with the call rate λ, a call rate is the average number of calls in a time unit,
- there are m servers with the same service rate μ, the service rate is the average number of calls that can be served in a time unit,
- each server can serve any call, one at a time,
- calls are Poisson with the rate λ,
- service times are exponential with the rate $\mu > 1/m\ \lambda$, in such a case
$$\lambda = 1/T_\lambda, \tag{15.1}$$
$$\mu = 1/T_\mu, \tag{15.2}$$
where T_λ is an average time between the calls and T_μ is an average service time,
- arriving calls enter the first available server, if all the servers are busy then the call takes a free waiting place,
- there are r waiting places, that means that a call waits, if there are not more than r other calls waiting, otherwise, the call disappears,
- waiting places are common for all the calls
- the system is stationary, meaning that the parameters μ and λ are constant,
- c_m is the server running cost (\$ per time unit),
- c_t is the customer time cost (\$ per time unit),
- c_p is the lost customer cost (\$ per lost call),
- the optimal number of servers $m = m(c, r)$ minimizes the total cost $C(c, m)$[1] per time unit, including the server running cost, the customer time cost, and the lost customer cost at fixed parameters r and $c = (c_s, c_t, c_p)$,

[1]Only two parameters c and m are in $C(c, m)$. The reason is uncertainty of c, therefore, the optimization of m is repeated several times at different "scenarios" c.

- results are presented as the waiting time distribution functions $F_\tau^{m,r}(t) = P^{m,r}\{\tau < t\}$, where $P^{m,r}\{\tau < t\}$ is the probability that the waiting time τ will be less than t at fixed number of servers m and waiting places r,
- a family of functions $F_\tau^{c,r}(t)$ is defined for different parameters c, r, these functions are presented in a simple format, assuming that the number of servers $m = m(c,r)$ is obtained by minimizing the total cost $C(c,m,r)$.

2. CALCULATION OF STATIONARY PROBABILITIES

Under these assumptions (Tijms, 1994), the probability of k calls in the system, including both, waiting calls and calls in services

$$P_k(\mu,\lambda) \quad = \quad \begin{cases} \frac{\rho^k}{k!} P_0(\mu,\lambda), & \text{if } 1 \le k \le m, \\ \frac{\rho^k}{m! m^{k-m}} P_0(\mu,\lambda), & \text{if } m < k \le m+r, \end{cases} \tag{15.3}$$

where $\rho = \lambda/\mu$. The no-calls probability

$$P_0(\mu,\lambda) = (\sum_{k=0}^{m-1} \frac{\rho^k}{k!} + \frac{m^m}{m!} \sum_{k=m}^{m+r} (\frac{\rho}{m})^k)^{-1} \tag{15.4}$$

follows from the condition

$$\sum_{k=0}^{m+r} P_k(\mu,\lambda) = 1. \tag{15.5}$$

The probability of losing a call at given numbers m, r, μ, λ is

$$P_{m+r}(\mu,\lambda) = \frac{\rho^{m+r}}{m! m^r} P_0(\mu,\lambda). \tag{15.6}$$

The average waiting time

$$T(m,r,\mu,\lambda) = \sum_{l=0}^{r-1} P_{m+l}(\mu,\lambda)\ t_l, \ r > 0, \tag{15.7}$$

where $P_k(\mu,\lambda)$ is defined by expression (15.3) and $t_l = \frac{l+1}{m\mu}$. The waiting time distribution functions

$$F_\tau^{m,r}(t) = P^{m,r}(\mu,\lambda)\{\tau < t\} = 1 - \Psi_\tau^{m,r}(t),$$

$$\Psi_\tau^{m,r}(t) = \sum_{k=m}^{m+r-1} P_k(\mu,\lambda)\ P_k\{\tau > t\}, \tag{15.8}$$

Here $P_k\{\tau > t\}$ denotes the probability that waiting time τ will exceed some fixed t under condition that there are k calls

$$P_k\{\tau > t\} = \sum_{s=0}^{k-m} q_s(t), \ k \geq m. \tag{15.9}$$

Here $q_s(t)$ is the probability that s calls will be served during the time t and

$$q_s(t) = e^{-m\mu t}\frac{(m\mu t)^s}{s!}, \ s = 0, 1, ..., r. \tag{15.10}$$

3. ASYMPTOTIC EXPRESSIONS

If the number r of waiting places is very large, one considers asymptotic expressions ($r \to \infty$, $\rho < m$) as a reasonable approximation. The asymptotic values are denoted by the index ∞. They are used to test the software designed for finite r. The no-calls probability

$$P_0^{\infty}(\mu, \lambda) = (\sum_{k=0}^{m-1} \frac{\rho^k}{k!} + \frac{\rho^m}{(m-1)!(m-\rho)})^{-1}. \tag{15.11}$$

The average waiting time

$$T(m, \infty, \mu, \lambda) = \frac{\rho^m}{m!m\mu(1-\rho/m)^2} P_0^{\infty}(\mu, \lambda). \tag{15.12}$$

The waiting time distribution functions

$$\begin{aligned} F_{m,\infty}(t) &= P_{m,\infty}(\mu, \lambda)\{\tau < t\} \\ &= 1 - \Psi_{m,\infty}(t), \end{aligned}$$

$$\Psi_{m,\infty}(t) = \frac{1}{1-\rho/m} e^{-(m\mu-\lambda)t} P_m^{\infty}(\mu, \lambda). \tag{15.13}$$

Here

$$P_m^{\infty}(\mu, \lambda) = \frac{\rho^m}{m!} P_0^{\infty}(\mu, \lambda). \tag{15.14}$$

The asymptotic probability of losing a call is zero.

4. "SURROGATE" SERVICES

An incoming call gets a regular service by an agent, if the actual or estimated waiting time is less then τ_w sec. Otherwise, a call gets a "surrogate" service by voice-mail. For simplicity, one replaces this service system by a system with $r = \tau_w/\tau_s$, where the expected service time is $\tau_s = 1/m\mu$. Then one can use all the expressions for the queuing system with r waiting places. Here, the call served by voice-mail is considered as the "lost" call. One considers calls of different type separately, because they are served by different agents.

5. CALL RATE ESTIMATE

The lost calls are not registered, as usual. Therefore, it is difficult to estimate the call rate λ directly, if r is limited. Then one uses least square estimates. The square deviations $\Delta(\mu, \lambda)$ between stationary probabilities $P_k(\mu, \lambda)$ and their estimates P_k^0 are minimized.

$$\Delta(\mu, \lambda) = \sum_{k=0}^{m+r} (P_k(\mu, \lambda) - P_k^0)^2. \qquad (15.15)$$

The stationary probabilities that there are k calls in the system are defined by expression

$$P_k(\mu, \lambda) \quad = \quad \begin{cases} \frac{\rho^k}{k!} P_0(\mu, \lambda), & \text{if } 1 \le k \le m \\ \frac{\rho^k}{m! m^{k-m}} P_0(\mu, \lambda), & \text{if } m \le k \le m + r. \end{cases} \qquad (15.16)$$

The least squares estimate of the call rate

$$\lambda^o = \arg\min_{\lambda} \Delta(\mu, \lambda). \qquad (15.17)$$

The estimates P_k^0 are obtained by counting the numbers of waiting calls at different time moments. Additional errors are expected, if the average number of waiting calls in a time unit is counted, instead of moment numbers.

6. OPTIMIZATION OF NUMBER OF SERVERS

The total cost of a service system

$$C(c, m) = c_m m + c_t T(m, r, \mu, \lambda) + c_p P_{m+r}(\mu, \lambda). \qquad (15.18)$$

One optimizes the server number m by simple comparison of different m within reasonable bounds $m_{min} \le m \le m_{max}$, $m_{min} > \lambda/\mu$

$$m(c, r) = arg \min_{m_{min} \le m \le m_{max}} C(c, m, r). \qquad (15.19)$$

7. MONTE CARLO SIMULATION (MCS)

The explicit steady-state solutions (15.6)-(15.13) are simple and exact, under the assumptions. However, there is no general explicit solution, if the call rate λ, and/or the service rate μ depends on time. That often happens in real service systems. The configuration and operating rules of real service systems are too complicated for the exact solution, as usual.

Then the Monte Carlo Simulation (MCS) is needed. We start the discussion of MCS from the steady-state systems keeping all the assumptions described in Section 1.2. That helps to test both the analytical and the Monte Carlo solutions by comparing the results. Later, MCS is extended to more complicated configurations and to time-dependent cases.

7.1 EVENT GENERATION

The events in a queuing system are the moments when a call arrives, when a call enters a server, and when a call leaves the system. To generate events, one needs two types of random number generators. The first type generates times until the next call. The second type generates service times.

Denote by $F_a(t) = P_a\{\tau < t\}$ the distribution function, where $P_a\{\tau < t\}$ is the probability that a random time τ will be less then t. Here a is the expected value of τ. Denote by $\xi \in [0,1]$ the random variable uniformly distributed between zero and one. Then

$$\tau = F^{-1}(\xi). \tag{15.20}$$

In exponential cases

$$F_a(t) = 1 - e^{-1/at}, \tag{15.21}$$

$$\tau = -aln(1 - \xi). \tag{15.22}$$

Here $a = 1/\lambda_s$, if times until the next call are generated.
$a = 1/(m_s \mu_s)$, if service times are generated.
The model defines moments when a call enters a server and leaves the system. These moments depend on the specific structure of the system. The model is designed trying to represent the actual operations as realistically as possible. However, the are limits that depend on the available computing power.

Here is a description of an algorithm of modeling and optimization of call centers:

- denote by t_n the "arriving" time of the nth call, then $t_{n+1} = t_n + \xi_n(\lambda)$, $\xi(\lambda) = -1/\lambda\ log(1-\eta)$, where η is a random number, uniform in $[0,1]$, and λ is the average number of calls in a time unit, the moment of the first call moment $t_1 = 0$ is the start of MCS,

- denote by τ_n the "departing" time of nth call, then $\tau_{n+1} = \tau_n + \xi_n(\mu_n)$, $\xi(\mu_n) = -1/\mu_n\ log(1-\eta)$, where μ_n is the average number of calls which can be served in a time unit by all $1 \leq m_n \leq m$ servers operating at the moment τ_n, thus $\mu_n = m_n\ \mu$, and μ is the average number of calls, served by single server in a unit time,

- count all calls in the system at the moment t_{n+1} including those in m servers and r waiting places by this expression

$$I_{n+1} = \begin{cases} I_n - 1, & \text{if } t_{n+1} > \tau_n \\ \min(I_n + 1, m + r), & \text{if } t_{n+1} < \tau_n, \end{cases} \tag{15.23}$$

  where I_n is the number of calls in the system at the moment t_n,

- define the number of servers m_n operating at the moment τ_n by this condition

$$m_n = \min(I_l, m),\ t_l \leq \tau_n \leq t_{l+1}, \tag{15.24}$$

- count calls lost up to the moment t_{n+1} by the expression

$$L_{n+1} = \begin{cases} L_n + 1, & \text{if } I_n + 1 > m + r, t_{n+1} < \tau_n \\ L_n, & \text{otherwise}, \end{cases} \tag{15.25}$$

  where L_n is the number of calls lost up to the the moment t_n,

- count the waiting time by this counter

$$T_{n+1} = T_n + \delta_n, \tag{15.26}$$

$$\delta_n = \tau_n - t_n, \tag{15.27}$$

- count the number of calls q^k that waited less than t^k, $t^{k+1} > t^k$, $k = 1, ..., K$

$$q^k_{n+1} = \begin{cases} \min(q^k_n + 1, m + r), & \text{if } \delta_n < t^k \\ q^k_n, & \text{if } \delta_n \geq t^k, \end{cases} \tag{15.28}$$

- define the distribution of the call numbers q^k that waited less than $t^k,\ t^{k+1} > t^k,\ k = 1, ..., K$

$$Q(t^k) = 1/N\ q_N^k, \tag{15.29}$$

- define the optimal number of servers m by simple comparison of different m within some reasonable bounds $m_{min} \leq m \leq m_{max}$

$$m(c,r) = arg \min_{m_{min} \leq m \leq m_{max}} C(c,m,r), \tag{15.30}$$

- compare the results with the corresponding steady-state explicit solutions, including the average waiting time, the average of lost calls and the waiting time distribution functions.

7.2 MONTE CARLO ERRORS

Under independence conditions, the standard deviation $\sigma(K)$ of the results $\theta(K)$ of a statistical model depends on the number of observations K as

$$\sigma(K) = \sigma(1)/\sqrt{K}. \tag{15.31}$$

Here $\sigma(1)$ is the "initial" standard deviation

$$\sigma(1) = \sqrt{E(\theta(1) - \Theta(1))^2}, \tag{15.32}$$

where $\Theta(1) = E\theta(1)$ and E is the expectation symbol.

In Monte Carlo simulation, $\Theta(1)$ cannot be estimated directly using the results $\theta(1)$ obtained by the 1-th observation. Many observations are needed, to reach a stationary state (one observation means one call). Therefore, we repeat the Monte Carlo simulation L times doing K observations each time. Then the error variance can be estimated as

$$\sigma_L^2(K) = 1/(L-1) \sum_{l=1}^{L} (\theta_l(K) - 1/L \sum_{j=1}^{L} \theta_j(K))^2. \tag{15.33}$$

Expression (15.33) is simpler, if the exact solution $\Theta = \lim_{K \to \infty} \theta(K)$ is known . Then

$$\sigma_L^2(K) = 1/(L-1) \sum_{l=1}^{L} (\theta_l(K) - \Theta)^2 \tag{15.34}$$

Therefore, to test a Monte Carlo procedure, one considers a model with exact solution, first. Then one applies MCS to more complicated models.

7.3 STOPPING MONTE CARLO

The Monte Carlo errors are important to define stopping rules. The traditional idea is that the computing errors should not be greater then the data gathering errors. In call centers, data errors depend on the errors of call rate predictions, as usual. Thus, the stopping rule is:
- stop at the first observation $k \geq K$ satisfying this inequality:

$$\sigma(K) < \sigma. \tag{15.35}$$

Here σ is a standard deviation of call rate predictions.

8. COMMON WAITING

The common waiting system is an example were one applies approximate models, including the Monte Carlo. By common waiting we define a system that serves calls of different type. One represents them as a call-vector $\lambda = (\lambda_1, ..., \lambda_l)$. Different calls are served by different servers. However, there are $r(c)$ of common waiting places. There are no simple formulas, such as (15.16) and (15.15). Therefore, we consider two approximate solutions: the analytical "reservation" model, and the Monte Carlo one.

8.1 ANALYTICAL APPROXIMATION: RESERVATION MODEL

One estimates $\lambda = (\lambda_1, ..., \lambda_l)$ by minimizing the square deviations

$$\Delta_s(\mu, \lambda, r_s) = \sum_{k_s=0}^{m_s+r_s} (P_{k_s}(\mu_s, \lambda_s, r_s) - P_{k_s}^0)^2. \tag{15.36}$$

Here $P_{k_s}(\mu_s, \lambda_s, r_s)$ is the stationary probability that there are k_s calls in the s-th server. This probability is defined by expressions similar to (15.3), assuming that there are r_s waiting places reserved for the calls λ_s. $r = (r_1, ..., r_l)$ is the reservation vector and $\sum_s r_s = r(c)$.

$$P_{k_s}(\mu_s, \lambda_s, r_s) = \begin{cases} \frac{\rho_s^k}{k_s!} P_0(\mu_s, \lambda_s, r_s), & \text{if } 1 \leq k_s \leq m_s, \\ \frac{\rho_s^k}{m_s! m_s^{k-m_s}} P_0(\mu_s, \lambda_s, r_s), & \text{if } m_s \leq k_s \leq m_s + r_s. \end{cases}$$

The estimates $P_{k_s}^0$ are obtained by counting the numbers of waiting s-calls at different time moments. The least square estimate of the call-vector λ is as follows

$$\lambda^o = \arg\min_{\lambda} \min_{r} (\sum_{s} \Delta(\mu_s, \lambda_s, r_s)). \tag{15.37}$$

The reservation model is simple and clear. However, one must test the reservation assumption by the Monte Carlo model.

8.2 STATISTICAL APPROXIMATION: MONTE CARLO MODEL

8.3 CALL RATE ESTIMATE

The statistical model can be used to estimate the call rates λ in the same way as the analytical one. One estimates $\lambda = (\lambda_1, ..., \lambda_l)$ by minimizing the square deviations $\Delta(\mu, \lambda, r)$ between the probabilities and their estimates

$$\Delta(\mu, \lambda, r) = \sum_{s} \sum_{k_s=0}^{m_s+r} (P_{k_s}(\mu, \lambda, r) - P^0_{k_s})^2. \tag{15.38}$$

Here $P_{k_s}(\mu, \lambda, r)$ is the probability[2] that there are k_s calls in the s-th server[3]. The estimates $P^0_{k_s}$ are obtained by counting the numbers of waiting s-calls at different time moments.

The least squares estimate of the call-vector λ

$$\lambda^o = \arg\min_{\lambda} \Delta(\mu, \lambda, r)). \tag{15.39}$$

The statistical model needs considerable computing power.

8.4 TESTING ANALYTICAL APPROXIMATION

The analytical approximation (see Chapter 8.1) is simpler. However, it is based on the reservation assumption. The statistical model may represent most of the important factors. However, it takes a long time

[2]Estimated using the Monte Carlo techniques described in the previous chapter.
[3]There are no reserved waiting places for the calls λ_s, thus the symbols r_s are omitted.

to "filter out" the random deviations. In the on-line operations, this time is too long, as usual.

To test analytical approximation, one compares estimates of call rates $\lambda = (\lambda_1, ..., \lambda_l)$ obtained by both the statistical and the reservation models. A small deviation means that the reservation model is acceptable. The large deviation means that the analytical approximation should be improved.

9. TIME-DEPENDANT CASES

Consider the MCS when the call rates $\lambda = \lambda(t)$, the number of servers $m = m(t)$, and the number of waiting places $r = r(t)$ depend on time. Therefore, all the results should be represented as functions of time. One modifies algorithms, similar to those in the previous chapter, by including the time factor:

- denote by t_n the "arriving" time of the nth call, then $t_{n+1} = t_n + \xi_n(\lambda(t)),\ \xi(\lambda(t)) = -1/\lambda(t)\ log(1-\eta)$, where η is a random number uniform in $[0,1]$ and λ is the average number of calls arriving in a time unit that includes the moment t,
- denote by τ_n the "departing" time of nth call, then $\tau_{n+1} = \tau_n + \xi_n(\mu_n),\ \xi(\mu_n) = -1/\mu_n\ log(1-\eta)$ where μ_n is the average number of calls which can be served in a time unit by all the $1 \leq m_n \leq m(\tau_n)$ servers operating at the moment τ_n, thus $\mu_n = m_n\ \mu$ and μ is the average number of calls served by single server in unit time,
- count the number of all the calls in the system at the moment t_{n+1}, including those in $m(t_{n+1})$ servers and $r(t_{n+1})$ waiting places

$$I_{n+1} = \begin{cases} I_n - 1, & \text{if } t_{n+1} > \tau_n \\ \min(I_n + 1, m(t_{n+1}) + r(t_{n+1})), & \text{if } t_{n+1} < \tau_n, \end{cases}$$

  where I_n is the number of calls in the system at the moment t_n,
- count the number of servers m_n operating at the moment τ_n

$$m_n = \min(I_l, m(\tau_n)),\ t_l \leq \tau_n \leq t_{l+1}, \tag{15.40}$$

- count the number of calls lost up to the moment t_{n+1}

$$L_{n+1} = \begin{cases} L_n + 1, & \text{if } I_n + 1 > m(t_{n+1}) + r(t_{n+1}), t_{n+1} < \tau_n \\ L_n, & \text{otherwise,} \end{cases}$$

where L_n is the number of calls lost up to the the moment t_n,

- define the waiting time

$$T_{n+1} = T_n + \delta_n, \tag{15.41}$$
$$\delta_n = \tau_n - t_n, \tag{15.42}$$

- count the number of calls s q^k that waited less than t^k, $t^{k+1} > t^k$, $k = 1, ..., K$

$$q_{n+1}^k = \begin{cases} \min(q_n^k + 1, m(t_{n+1}) + r(t_{n+1})), & \text{if } \delta_n < t^k \\ q_n^k, & \text{if } \delta_n \geq t^k, \end{cases}$$

- define the distribution of the call numbers q^k that waited less than t^k, $t^{k+1} > t^k$, $k = 1, ..., K$

$$Q(t^k) = 1/N \ q_N^k, \tag{15.43}$$

- optimize the number of servers $m(t)$ at different times t by solving the corresponding stochastic scheduling problem[4]

- define the difference between the steady-state and dynamic solutions by comparing the average results with the corresponding steady-state explicit solutions, including the waiting time distribution functions $F_\tau^{m,r}(t) = P^{m,r}\{\tau < t\}$, where $P^{m,r}\{\tau < t\}$ is the probability that the waiting time τ will be less than t,

- define the iteration number N by Monte Carlo experimentation.

9.1 SIMPLE EXAMPLE

Consider this simple example

$$\lambda(t) = a + b \ sin(\phi + 2\pi\omega t). \tag{15.44}$$

Here $\omega = 1/24$ and $\phi = 8/24$. t is caunted in hours. a is average number of calls in 24 hours. b is amplitude of daily variation of calls, $r(t) = r$.

$$m(t) = \begin{cases} m^0, & \text{if } sin(\phi + 2\pi\omega t) < 0, \\ m^1, & \text{if } sin(\phi + 2\pi\omega t) > 0. \end{cases} \tag{15.45}$$

[4]Solution of this problem is difficult, thus, later on we shall consider some simple cases as illustrations.

One optimizes the numbers m^0 and m^1 of two different servers considering two periods. This is a short description of the optimization problem:

- there are two optimization parameters m^0 and m^1 restricted by bounds $m_{min} \leq m^0 \leq m^0_{max}$ and $m^1_{min} \leq m^1 \leq m^1_{max}$,
- one may obtain the optimal pair by direct comparison of all the feasible pairs

$$m* = (m^{*0}, m^{*1}) = arg \min_{m^0, m^1} C(c^0, c^1, m^0, m^1, r), \qquad (15.46)$$

- the total system cost is defined as

$$C(c^0, c^1, m^0, m^1, r) = c^0 m^0 + c^1 m^1 + c_t T(m^0, m^1, r, \mu, a, b, \phi, \omega) + c_p P_{m^0, m^1, r}(\mu, a, b, \phi, \omega). \qquad (15.47)$$

10. CALL RATE PREDICTIONS

10.1 INTRODUCTION

Consider ideas and models that supplement the ones described in chapter 14. The call rate depends on many factors (see Figures 15.1 and 15.2). Therefore, one applies the extended version of the Auto-Regression-Moving-Average (ARMA) model and software (see Chapter 14).

The results show that one does not improve predictions by including the external factors directly into the ARMA model. A reason is that in the ARMA framework it is difficult to estimate the delay time (see expression (14.22)) and the duration of impacts Special Events (SE)[5]. The additional difficulty is that most of SE are rare (see Figure 15.1). Therefore, columns of date file are filled mostly by zeros. Most of the SE indicated in Figure 15.2 are predictable. For example, one can predict factor 6 (public holidays, traditional celebrations) and factor 7 (last day for ordering) exactly. Other SE can be predicted approximately. The knowledge of future values of external factors helps to predict the main

[5]While predicting call rates, we call the external factors as "Special Events" (SE).

1	10/1/96	Tue		1607	1607	1.2	1.2
0	0	0	0	0	0	0	0
2	10/2/96	Wed		1431	1431	1.1	1.1
0	0	1	0	0	0	0	0
3	10/3/96	Thu		1400	1400	1.1	1.1
0	0	0	0	0	0	0	0
4	10/4/96	Fri		1194	1194	1.1	1.1
0	0	0	0	0	0	0	0
5	10/5/96	Sat		476	476	1.1	1.1
0	0	0	0	0	0	0	0
6	10/7/96	Mon		1553	1553	1	1
0	1	0	0	0	0	0	0
7	10/8/96	Tue		1421	1421	1	1
0	0	0	0	0	0	0	0
8	10/9/96	Wed		1486	1486	1.1	1.1
0	0	0	0	0	0	0	0
9	10/10/96	Thu		1339	1339	1.1	1.1
0	0	0	0	0	0	0	0
10	10/11/96	Fri		1106	1106	0.86	0.86
0	0	0	0	0	0	0	0

Figure 15.1. A fragment of the the DATAFILE 'call.data' including the number, the date, the call rate (repeated twice) , the real-valued external factor (repeated twice) and the indicators of eight Boolean external factors.

Events

index	code	description
1	Cp	Postage of main catalogue
2	Cr	Members receive main catalogue
3	D	Selecting members for dunning
4	Ep	Postage of extra catalogue
5	Er	Members receive extra catalogue
6	H	Public Holidays, traditional celebrations
7	L	Last day for ordering
8	Op	Postage of Order-reminder

Figure 15.2. Eight Boolean external factors related to call rate.

one. However, the present software version cannot use this possibility yet.

Here, we consider call rates $\lambda(t)$ as a sum of two stochastic functions

$$\lambda(t) = z(t) + \nu(t). \tag{15.48}$$

Often the rate $\lambda(t)$ has several components $\lambda(t) = (\lambda_1(t), ..., \lambda_l(t))$ corresponding to different types of calls. In expression (15.48) $\nu(t)$ denotes a "stationary" component which is described by the ARMA model. A "non-stationary" component is denoted by $z(t)$. This way we separate the stationary part from the non-stationary one.

The theoretical analysis of non-stationary stochastic functions is difficult. Therefore, the separation of non-stationary component is important. The non-stationary component is defined by a local expert, as usual. The expert applies his knowledge while using the previous data and making the future predictions. Thus, we call $z(t)$ as the "expert" component and $\nu(t)$ as the "statistical" one. The estimate of the statistical component $\nu(t)$ is investigated in Chapter 14, considering the ARMA model. An alternative is "Scale" models. They predict the expert component $z(t)$ by estimated scales. The scales express the differences between different SE and different times.

10.2 CALL RATE PREDICTION BY SCALE MODELS

The ARMA models considered in Chapter 10. reflect non-stationarity by eliminating unstable parameters applying the structural stabilization techniques. This way some data and some model parameters a_i, b_j are eliminated. The estimates of the remaining ones are obtained. Therefore, one may regard ARMA parameters a_i, b_j as some scales, too. These scales reflect the influence of the corresponding data sets by minimization of prediction errors. No expert knowledge is involved.

Now we shall consider the scale models, where the expert opinion is involved. This is done by choosing data sets and scales reflecting the expert opinion. Two versions of scales models are considered.

In the first model it is supposed that the prediction z_i is a product of the present call rate z_{i-1} and some scale s_i

$$z_i = z_{i-1}\ s_i. \tag{15.49}$$

Here z_i is a predicted call rate, called a "prediction." z_{i-1} is an observed call rate, called a "data set." The index $i-1$ defines the data set used to predict z_i. The parameter s_i is the "scale" to predict z_i by the data set z_{i-1}. Expression (15.49) is a time scale model. In this model the scale is estimated as

$$s_i = \frac{z_{p(i)}}{z_{p(i-1)}}. \tag{15.50}$$

In (15.50) the subscript $p(i)$ defines a period which precedes i. The subscript $p(i-1)$ denotes a period preceding $i-1$. The term "preceding" means the nearest previous period of the same type. For example, Saturdays, Sundays, holidays, Christmas weeks, sporting events, days of marketing messages, e.t.c., all are different types.

In the second model

$$z_i = z_{p(i)}\ s_{i-1}. \tag{15.51}$$

Here the scale is estimated as

$$s_{p(i)} = \frac{z_{i-1}}{z_{p(i-1)}}. \tag{15.52}$$

Both expressions(15.49) and (15.51) predicts the same call rates. That means that one obtains the same results by the event scales (15.50) and the time scales (15.52). Only the interpretation differs, in scalar cases.

In the vector case, expressions (15.49) and (15.51) predict different call rate graphs. One represents these graphs as vectors $z_i = (z_{ij},\ j = 1, ..., J)$. The components z_{ij} denote call rates of different parts j of a period i. For example, hours if we predict next day call rates.

The assumption of the first model (15.49) is that the next day graph $z_i = (z_{ij},\ j = 1, ..., J)$ is equal to the present one $z_{i-1} = (z_{ij},\ j = 1, ..., J)$ multiplied by scales s_i

$$s_i = \frac{Z_{p(i)}}{Z_{p(i-1)}}. \tag{15.53}$$

Here $Z_i = 1/J \sum_{j=1}^{J} z_{ij}$ denotes the average call rates of the period i. Expression (15.53) means that the shapes of the next and the present graphs remains the same. Only the scales differ.

The second model (15.51) assumes that the next day graph is equal to the graph of the preceding period $z_{p(i)} = (z_{p(i),j},\ j = 1, ..., J)$ multiplied by scales

$$s_{p(i)} = \frac{Z_{i-1}}{Z_{p(i-1)}}. \tag{15.54}$$

This means that the shapes of the next graph and the preceding one are the same. Only the scales differ. The numerator $Z_{p(i)}$ of scales s_i of the first model depends on the Special Events (SE) of the next period i. We call the first model (15.49) as the event scale model.

The numerator $Z_{p(i)}$ of scales s_i of the first model depends on the Special Events (SE) of the next period i. By definition, the preceding period $p(i)$ depends on the type of next one. The period type is defined by the Special Event (SE) which is active in the period. Examples of SE are Saturdays, Sundays, holidays, Christmas weeks, sporting events, days of marketing messages e.t.c..

The numerator Z_{i-1} of scales $s_{p(i)}$ of the second model depends on the average call rate now $i - 1$. Therefore, we call the second model (15.51) as the time scale model.

Using the event scale model (15.49), the next day event SE defines the next day scale. The shape of the next day graph remains the same as today. Therefore, this model reflects the changes of the graph shapes without delay.

Using the time scale model (15.51), the shape of the next day graph is the same as the shape of the preceding one. Here the graph changes, but with some delay. The interval between the next period and the previous period of the same type defines the delay time.

10.3 EXPERT MODEL, EVENT SCALE VERSION

DEFINITION OF SPECIAL EVENTS

Here the call rate in the next period (15.49) is the same as in the present one multiplied by the scale s_i. The scale depends on the Special Event (SE) which is expected to be active in the next period. The periods where no special events are active, we consider as periods with "neutral" special events. Scales s_i determine the impact of SE to the call rate z_i of next period.

The scale s_i reflecting the impact of a special event is estimated using empirical data by expressions (15.50), (15.53) and/or directly by expert opinions. The impact of some special events, such as Saturdays, Sundays, holidays, Christmas week e.t.c.. is well defined and instant. One calls them Instant Special Events (ISE).

The impact of some others, such as marketing messages, is delayed. Call rates react to these events with some delay. Therefore, one calls them Delay Special Events (DSE). The starts and the durations of DSE are fixed. The delay times d and the durations τ of DSE impacts are estimated using the observed data and expert knowledge. In some periods several DSE may be active simultaneously. One calls these Multiple Special Event (MSE).

The starts and the durations of ISE and their impacts are equal to starts and durations of the corresponding periods. Beginnings and durations of DSE impacts are not necessarily equal to the beginnings and durations of the periods. Therefore, a special sorts of DSE appear, called Partial Special Events (PSE). In PSE the impact of a special event is active only during some part of the period. The delay times d, the durations τ of the impacts of DSE, and the scales s defining the impact of all the special events, are estimated using the available data and the expert knowledge.

PROCEDURES

We consider three expert procedures and the method of least squares, while defining estimates of scales s_i. The first procedure is the empirical one. The scales s describing the impact of all sorts of events, including the partial and the multiple ones, are estimated by expressions (15.50) or (15.53). These expressions define the relation of the call rates in the pair of periods $p(i-1)$ and $p(i)$. A period $p(i-1)$ precedes the present period $i-1$. Denote by $p(i)$ a period that precedes the next one, denoted i. The numerator $z_{p(i)}$ is the call rate[6] of the period preceding the next

[6] In vector predictions that is an average call rate $Z_{p(i)}$.

one. The denominator $z_{p(i-1)}$ is the call rate of the period preceding the present one.

Using the empirical procedure, partial and multiple special events are considered as different SE. Here scales s are defined separately for each of them. Therefore, the empirical procedure is convenient, if the data is available for all types of periods, including the partial and multiple ones. In such a case, estimate of the delays d and durations τ of DSE is not needed. It is supposed that an identical situation[7] is in the available data. One needs large data sets for that.

The second direct procedure is the subjective one. All unknown parameters of the expert model, such as the scales s, the delay times d and the duration times τ, are defined by the expert opinion. This is a reasonable way to start a new system, when no data is available.

The third direct procedure is a mixture of the empirical and the subjective ones. We use empirical estimates, if the corresponding data is available[8]. Otherwise, one uses subjective estimates. Using the method of least squares, we search for such scales s, such delay and duration times d, t that minimize the sum of squared deviations. The deviations are differences between the call rates predicted by the expert model and the observed call rates z_{ij}. All four procedures are applied to different times: hours, days, weeks and seasons. First, we consider the scalar case.

SCALAR PREDICTION

In the scalar case, one predicts a single number, the average call rate of the next period. In the formal terms, one considers the call rate function z_i as a sample of some non-stationary stochastic process. Four different periods; hours, days, weeks, and seasons, are considered independently. Let us start by describing hours.

---

[7]That means, there exist periods in the data file, with the identical combinations of impacts. Besides, these combinations are the same as that of the new and the present periods.

[8]If a pair of periods, identical to the pair of the present and the next one, can be found.

Hourly Prediction. Using the event scale model, a call rate z_i of the next hour i is expressed this way

$$z_i = z_{i-1} \ s_i. \tag{15.55}$$

Here z_{i-1} is the present call rate. s_i is a scale of the next hour i. The hourly scales s_i differ depending on the impact of SE felt at the day i. For example, working days, weekends, Christmas days, marketing messages days, e.t.c.

$$s_i = \frac{z_{p(i)}}{z_{p(i-1)}}. \tag{15.56}$$

Here $p(i)$ is a subscript of the hour preceding the next one. $p(i-1)$ is a subscript of the hour preceding the present one. Following examples illustrate what one means by the term "preceding."

Examples of Multiple Special Events (MSE). Suppose that the impact of the special event "marketing messages send" starts at the next hour i, that is the second one of a working day. Assume that the present hour $i-1$ is the first one of a working day. Then

$$s_i = \frac{z_{p(i)}}{z_{p(i-1)}}. \tag{15.57}$$

Here $p(i)$ is the subscript of the most recent previous second hour of a working day with the same impact SE. $p(i-1)$ is the subscript of the most recent previous first hour of a working day with no SE. In this example, we consider a multiple SE as two single SE. The first single SE means the first working hour without the impact of marketing messages. The second single SE defines the second working hour with the impact of marketing messages.

Another way is to consider the impacts of those two SE separately. Then one defines the scale s_i as a product of two scales

$$s_i = s_i^1 \ s_i^2. \tag{15.58}$$

Here

$$s_i^1 = \frac{z_{p^1(i)}}{z_{p^1(i-1)}}, \tag{15.59}$$

where $p^1(i)$ is the index of the first hour and $p^1(i-1)$ is the index of the second hour. Both indices denote the previous working day.

$$s_i^2 = \frac{z_{p^2(i)}}{z_{p^2(i-1)}}, \tag{15.60}$$

where $p^2(i)$ is the index of the most recent hour with "marketing" and $p^1(i-1)$ is the index the most recent hour with no "marketing." Expression (15.58) needs less data. However, it is based on the assumption that scales are multiplicative. Expression (15.58) involves expert opinion by assuming multiplicativity of scales. Expression (15.57) is an example of empirical approach; multiple SE are considered as different special events.

Examples of Partial Special Events (PSE). The only difference from the previous example is that the impact of "marketing messages" starts at the 20-th minute of the next hour i. Supose that PSE is a special sort of SE with the scale

$$s_i = \frac{z_{p(i)}}{z_{p(i-1)}}. \tag{15.61}$$

Here $p(i)$ is the subscript of the most recent previous day when the "marketing" starts at the 20-th minute of second working hour. $p(i-1)$ is the subscript of the most recent previous first hour of a working day with no marketing. Another way is to consider the impact of PSE by setting the scale s_i of marketing messages, in proportion to the duration of their impact

$$s_i^{20} = 2/3\ s_i, \tag{15.62}$$

where

$$s_i = \frac{z_{p(i)}}{z_{p(i-1)}}. \tag{15.63}$$

Here $p(i)$ is the subscript of the most recent previous day when the "marketing" starts at the second working hour. $p(i-1)$ is the subscript of the most recent previous first hour of a working day with no marketing.

Expression (15.62) needs less data. However, it is based on the assumption that scales are proportional to the impact duration. Expression (15.62) involves expert opinion by assuming the proportionality of scales. Expression (15.61) is an example of the empirical approach where a PSE is considered as a different special event.

METHOD OF LEAST SQUARES

Denote by $\psi_i(s, d, \tau)$ the call rate prediction by an expert model with fixed parameters s, d, τ. One predicts call rates of the next period i, using data up to the period $i-1$. The parameters of the model include scales s and delay and duration times d, τ. We minimize the sum

$$\min_{s,d,\tau} \sum_{i=T_0}^{T} (z_i - \psi_i(s,d,\tau))^2. \tag{15.64}$$

Here T is the end and T_0 is the beginning of the "learning" set.

One minimizes (15.64) by various global and local optimization methods. That means that estimates of parameters of expert scale models are obtained in the same way as estimates of ARMA parameters (see Section 14). The difference of (15.64), is that parameters s, d, τ of the expert model EM are optimized, instead of ARMA parameters b. The optimization problem (15.64) is very difficult. Here, the number of variables is great and the objective function is multi-modal.

VECTOR PREDICTION, EVENT SCALE VERSION

One often predicts the graph of call rates of the next perid. The graph is represented as call rates of different parts of the next period. These call rates are components of some vector. That justifies the term "vector prediction"[9].

The only difference from the scalar prediction is that we predict not a single call rate z_i of the next period i, but some vector $z_i = (z_{ij},\ j = 1, ..., J)$. Here the component z_{ij} definines the call rate of the part j of the period i. First, we consider daily predictions, where index i denotes days and the index j denotes hours.

Daily Vector Prediction. We predict the hourly call rates for the next day. The predicted call rate z_{ij} of the hour j of the day i is

$$z_{ij} = z_{i-1,j}\ s_i. \tag{15.65}$$

Here $z_{i-1,j}$ is the call rate of the hour j of the present day $i-1$. s_i is a scale of the next day i. The scales s_i depend on the impact of SE felt at the day i. For example, working day, Saturday, Sunday, Christmas day,

[9] In section 10. the terms "one-step" and "multi-step" were used instead of "scalar" and "vector."

marketing day, e.t.c.

$$s_i = \frac{Z_{p(i)}}{Z_{p(i-1)}}. \tag{15.66}$$

Denote by $p(i)$ the subscript of the nearest previous day, similar to the next one. Denote by $p(i-1)$ the subscript of the nearest previous day, similar to the present one. The meaning of "nearest previous day similar to" is illustrated by following examples.

Examples of MSE. Suppose that the impact of the SE "marketing messages send" starts at the next day i. That is Tuesday. The present day $i-1$ is Monday with no additional special events. Then the scale

$$s_i = \frac{Z_{p(i)}}{Z_{p(i-1)}}. \tag{15.67}$$

Here $p(i)$ is the subscript of the most recent previous Tuesday with the same SE. $p(i-1)$ is the subscript of the most recent previous Monday with no SE. In this example we considered a multiple SE as two single SE: Monday without marketing, and Tuesday with marketing.

Another way is to consider the impacts of those two SE by defining the scale s_i as a product of two scales.

$$s_i = s_i^1 \; s_i^2 \tag{15.68}$$

Here

$$s_i^1 = \frac{Z_{p^1(i)}}{Z_{p^1(i-1)}}, \tag{15.69}$$

where $p^1(i)$ is the index of Monday, $p^1(i-1)$ is the index of Tuesday of the previous week, and

$$s_i^2 = \frac{Z_{p^2(i)}}{Zz_{p^2(i-1)}}. \tag{15.70}$$

$p^2(i)$ is the index of the most recent day with "marketing." Index $p^1(i-1)$ denotes the most recent day with no "marketing." Expression (15.67) is an example of empirical approach. Here multiple SE are regarded as different special events. Expression (15.68) needs less data but is based on an expert assumption that scales are multiplicative.

Examples of PSE. One feels the impact of the special event, called "marketing," next day i on the sixth working hour. That is the only difference from the previous example. Let us to consider PSE as another SE. Then

$$s_i = \frac{Z_{p(i)}}{Z_{p(i-1)}}. \tag{15.71}$$

Here $p(i)$ is the subscript of the most recent previous week when the "marketing" starts Tuesday at the sixth working hour. $p(i-1)$ is the subscript of the most recent previous Monday with no marketing.

Another way is to regard PSE as a reduced SE. Then, the scale s_i is proportional to the impact duration

$$s_i^6 = 1/4\ s_i. \tag{15.72}$$

Here

$$s_i = \frac{Z_{p(i)}}{Zz_{p(i-1)}}, \tag{15.73}$$

$p(i)$ is the subscript of the most recent previous week, such that "marketing" starts Tuesday on the sixth hour, and $p(i-1)$ is the subscript of the most recent previous Monday with no marketing.

Expression (15.71) is an example of empirical approach when a PSE is regarded as a different special event. Expression (15.72) needs less data. However, it is based on an expert assumption that scales are proportional to the impact duration.

METHOD OF LEAST SQUARES

Denote by $\psi_{ij}(s,d,\tau)$ the call rate prediction by an expert model with fixed parameters s, d, τ. One predicts call rates of the part j of the next period i using data up to the period $i-1$. The parameters of the expert model include scales s and delay and duration times d, τ. All special events, including DSE and PSE, are regarded. We minimize the sum

$$\min_{s,d,\tau} \sum_{i=T_0}^{T} \sum_{j=1}^{J} (z_{ij} - \psi_{ij}(s,d,\tau))^2. \tag{15.74}$$

Here T is the end and T_0 is the beginning of the "learning" set.

Define the expression (15.74) as a function 'fi' in the file 'fi.C'. Then one minimizes (15.74) by various global and local optimization methods. That means that estimates of parameters of expert scale models are obtained in the same way as estimates of ARMA parameters (see Section 14).

10.4 TIME SCALE VERSION, VECTOR PREDICTION

Scalar predictions by time scales are identical to those using event scales. Vector predictions by time scale models (15.51) are different from those using event scale models (15.49). Here the shape $z_I = (z_{ij},\ j = 1, ..., J)$ of the next day graph remains the same as that on the preceding day $p(i)$. Only scales s_{i-1} change. This follows from (15.52). Scales s_{i-1} of the present period $i-1$ reflect changes of average call rates between the present period and the preceding one. The scales are estimated using empirical data and/or expert opinion. The definitions and examples of all the special events are the same as in the Event Scale case (see Section 10.3).

PROCEDURES

We shall consider four procedures to obtain estimates: empirical, subjective, mixed, and least squares. The estimates are similar to those in the Event Scale Model (see Section 10.3). Only scales and preceding periods are different. One predicts a vector $z_i = (z_{ij},\ j = 1, ..., J)$. Here z_{ij} means the call rate of the part j of the period i. First, we consider daily predictions. Then the index i denotes days, and the index j denotes hours.

DAILY VECTOR PREDICTION

We predict the hourly call rates of the next day. The predicted call rate z_{ij} of the hour j of the next day i is

$$z_{ij} = z_{p(i)j}\ s_{i-1}. \tag{15.75}$$

Here $z_{p(i)j}$ is the call rate of the hour j of the day preceding the next one. s_{i-1} is a scale of today $i-1$. Scales s_{i-1} depend on the impact of SE felt at the day i. For example, working day, Saturday, Sunday, Christmas day, marketing day, e.t.c.

$$s_i = \frac{Z_{i-1}}{Z_{p(i-1)}}. \tag{15.76}$$

Here $p(i-1)$ is a subscript of the day preceding the present one. The term "preceding" is illustrated by the following examples.

Examples of MSE. Suppose that the impact of the SE "marketing messages send" begins and ends today $i-1$, which is Monday. The next day i is Tuesday with no additional special events. Then the scale

$$s_i = \frac{Z_{i-1)}}{Z_{p(i-1)}}. \tag{15.77}$$

The index $p(i-1)$ denotes the day, which precedes the present one. Here, we regard a multiple SE as two single SE. First single SE is Monday with marketing. The second single SE is Tuesday without marketing.

Another way is to regard the impacts of those two SE by defining a scale s_{i-1} as a product of two scales.

$$s_{i-1} = s_{i-1}^1 \; s_{i-1}^2. \tag{15.78}$$

Here

$$s_{i-1}^1 = \frac{z_{p^1(i-1)}}{z_{p(p^1(i-1))}}. \tag{15.79}$$

$p^1(i-1)$ is the index of the day preceding Monday without marketing. $p(p^1(i-1))$ is the index of the day preceding $p^1(i-1)$.

$$s_i^2 = \frac{z_{p^2(i-1)}}{z_{p(p^2(i-1))}}. \tag{15.80}$$

$p^2(i-1)$ is the index of the day preceding the day with marketing. $(p(p^2(i-1))$ is the index of the day preceding $p^2(i-1)$.

Expression (15.77) is an example of empirical approach when multiple SE are regarded as different special events. Expression (15.78) needs less data but is based on the assumption that scales are multiplicative.

Examples of PSE. Here, the impact of "marketing" starts at the second working hour of the present Monday $i-1$. Regarding PSE as another SE the scale is

$$s_{i-1} = \frac{z_{i-1}}{z_{p(i-1)}}. \tag{15.81}$$

Here $p(i-1)$ is the index of the day preceding the present one.

Another way is to regard the PSE as a reduced SE. Then the scale s_{i-1} is proportional to the impact duration

$$s_{i-1}^2 = 1/4 \ s_{i-1}. \tag{15.82}$$

Here

$$s_{i-1} = \frac{z_{p^3(i-1)}}{z_{p(p^3(i-1))}}. \tag{15.83}$$

$p^3(i-1)$ is the index of the preceding marketing Monday. $p(p^3(i-1))$ is the index of the day preceding $p^3(i-1)$. Applying expression (15.82) one requires less data. However, this expression is based on the assumption that scales are proportional to the impact duration.

Expression (15.82) involves expert opinion indirectly, by assuming proportional scales. Expression (15.81) is an example of empirical approach when a PSE is regarded as another SE.

METHOD OF LEAST SQUARES

Denote by $\psi_{ij}(s,d,\tau)$ the call rate prediction by an expert model with fixed parameters s,d,τ. One predicts call rates of the part j of the next period i using data up to the period $i-1$. The parameters of the expert model include scales s and delay and duration times d,τ. All special events, including DSE and PSE are regarded. We minimize the sum

$$\min_{s,d,\tau} \sum_{i=T_0}^{T} \sum_{j=1}^{J} (z_{ij} - \psi_{ij}(s,d,\tau))^2. \tag{15.84}$$

Here T is the end and T_0 is the beginning of the "learning" set. One minimizes (15.64) by corresponding global and local optimization methods.

REMARKS

As usual, expert models predict the weekly and the seasonal graphs. The weekly graphs are given in a form of average calls for each of seven days. In the seasonal graphs, weeks are represented by weekly averages. Often "observed" values of the call rates z_i cannot be defined directly. Then their estimates are used (see Section 5.). Missing or uncertain data is replaced by expert estimates. This is important for the first year of operation.

10.5 TIME SERIES MODEL (ARMA)

SCALAR PREDICTION

Here we consider how to apply the ARMA model to predict the errors ν_i of the expert scale model $\psi_i(s, d, \tau)$, where $\nu_i = \psi_i(s, d, \tau) - z_i$ (Mockus et al., 1997). The formal description of the ARMA model is in Chapter 14. The software is on the web-site (see Section 4.). If predicting ν_i we do not know some values of ν_s, $s = i-1, i-2, ...$ then we replace the missing data by the expected values of the unknown ν_s defined recurrently (see Section 10.).

ARMA models predict only the difference between the expert models (EM) and the data. Therefore, in the first step, EM models are adapted to the data by defining the right scales. Only then, the parameters a, b of ARMA models are optimized. This means that one minimizes a squared difference between the adapted EM model and the observed data. Predicting the data, the results of both EM and AH models are summed up.

VECTOR PREDICTION

The vector prediction of EM errors ν_i is performed using multi-step ARMA predictions described in Section 9. "Confidence cones" could be useful for preliminary evaluation of prediction errors. They estimate probable deviations from expected values with some "confidence level," say ninety percent or ninety nine percent. These cones are defined empirically as upper and lower limits of the results obtained by repetition of Monte Carlo procedures (see Section 9.).

10.6 APPLICATION EXAMPLES

In the call center example, both the Expert and ARMA models were tested separately. Figure 14.18 shows daily call rates of this center. Table 14.1 illustrates some results of the ARMA model. Results of the expert model are not available, for publication.

11. CALL CENTER SCHEDULING

There are "Of-the-Shelf" tools for the scheduling of single-skill agents. The scheduling of multi-skill agents is theoretically possible using Monte Carlo simulation (see Section 7.). However, the simulation time is too large for on-line scheduling. A convenient approximation to multi-skill scheduling is the reduction of multi-skill problems to the single-skill ones. One can do that by representing each multi-skill agent as a "weighted" single-skill one

$$N_{single} = v * N_{multi}. \tag{15.85}$$

Then, one estimates the unknown weight v by minimizing the sum of squared deviations. The deviations are differences between results of approximate single-skill model and the multi-skill one

$$\min_v 1/K \sum_{k=1}^{K} (Q_k(\lambda, N_{multi}) - q_k(\lambda, N_{single}))^2. \tag{15.86}$$

Here Q_k denotes the results of the k iteration of a Monte Carlo simulation of the multi-skill system. λ is the call rate. N_{multi} is the number of multi-skill agents. q_k is the result of the k iteration of a Monte Carlo simulation of a single-skill system. N_{single} is the number of single-skill agents, which replace the multi-skill ones.

Another way is to extend the single-skill simulation (see Section 11.) to the multi-skill case. Here one needs more computing time to estimate the "optimal" weights v for different call rates. For example, the method *Exkor* is used to obtain the best values of v (see Section 4.).

Chapter 16

OPTIMAL SCHEDULING

1. INTRODUCTION

Industrial, financial, commercial or any kinds of activity have at least one common feature: the better organized they are, the higher the profit, the better the quality or the lower the cost. Everyone makes a schedule to organize his or her own life. Making a schedule for any organized activities, one considers a sequence of tasks and a list of resources. Resources include tools, machines, materials, work force e.t.c. Of course, making a schedule for organization is far more difficult than making a schedule for our everyday life.

A failure to make a schedule or devising a wrong schedule can result in delay of a deadline and can cost much of the money. An example of early formal scheduling technique is the Harmonygraph developed in 1931 by Karol Adamiecki (Adamiecki, 1931). The review of network based scheduling techniques is in (Hajdu, 1997).

First we consider an example of a simple scheduling problem. Here the sequence of tools is fixed by technology. One minimizes the make-span, which is the time from the beginning of the first task until the end of the last one.

The second example is a school scheduling problem. Here the sequence of teaching subjects, regarded as tools, can be changed. One needs to reduce the sum of "empty" hours for teachers. There should be no empty hours for classes. Different classes are considered as different tasks. The classrooms, including the computer and physics rooms and studies, are the limited resources.

2. FLOW-SHOP PROBLEM

The flow-shop problem is a simple case of the large and important family of scheduling problems. We denote the sets of jobs and machines by J and S. Denote by $\tau_{j,s}$ the duration of operation (j, s). Here $j \in J$ denotes a job and $s \in S$ denotes a machine[1].

Suppose that the sequence of machines s is fixed for each job j. One machine can do only one job at a time. Several machines cannot do the same job at the same moment. The decision $d_i(j) \in D_i$ means starting the job $j \in J_i$ at the stage i. We define the set of feasible decisions D_i as the set J_i of all the jobs available at the stage i and conforming to the flow-shop rules.

The objective function v is the make-span[2]. Denote by $T_j(d)$ the time when we complete the job j (including the gaps between operations) using the decision sequence $d = (d_1, ..., d_K)$, where K the number of decision stages. Then the make-span of this d is

$$v(d) = \max_{j \in J} T_j(d). \tag{16.1}$$

2.1 PERMUTATION SCHEDULE

The number of feasible decisions for the flow-shop can be very large. One reduces this number, if one considers only the so-called permutation schedules that are subsets of all feasible schedules. The permutation schedule is a schedule with the same job order on all machines. Here one defines a permutation schedule by fixing a sequence of job indices, for example, $1, 2, ..., n$. The schedule is transformed by a single permutation of job indices. As usual, permutation schedules describe the optimal decision well and are easier to implement (Baker, 1974).

[1]Describing the scheduling problems we use the terms machine and tool and the terms job and task as synonyms.
[2]The time to complete all the jobs.

Denote

$$\tau_j = \sum_{s=1}^{|S|} \tau_{j,s},$$

where $\mid S \mid$ stands for the number of machines. We define τ_j as the length of the job j.

2.2 HEURISTICS

Define Longer-Job heuristics

$$h_i(j) = \frac{\tau_j - A_i}{A^i - A_i} + a. \tag{16.2}$$

Here

$$A_i = \min_{j \in J_i} \tau_j, \ A^i = \max_{j \in J_i} \tau_j, \ a > 0. \tag{16.3}$$

The priority rule (16.2) prefer a longer job.

2.3 RESULTS

Table 16.1 illustrates results of BHA after 100 iterations using the Longer-Job heuristic (16.2) and different randomization procedures. The objective is a stochastic function. This function defines the shortest make-span found after K repetitions.

In the example, $J = S = O = 10$, where J, S, O are numbers of jobs, machines, and operations, respectively. Lengths and sequences of operations are generated as random numbers uniformly distributed from zero to ninety-nine. The expectations and standard deviations are estimated by repeating forty times the optimization of a randomly generated problem. In Table 16.1 the symbol f_B denotes a mean, and d_B denotes a standard deviation of the make-span. "Delta" denotes randomization including terms $l = 0, 1, 2, \infty$ (see expressions (2.6) and (2.7)). "Taylor 3" denotes randomization (2.6) where the number of terms is $L = 3$. "CPLEX" denotes the results of the well known general discrete optimization software after twenty hundred iterations (one CPLEX iteration is comparable to a Bayesian observation). The bad results of CPLEX show that the standard MILP technique is not efficient in solving this specific problem of discrete optimization. It is not yet clear how much one improve the results using specifically tailored B&B.

Table 16.1. The results of Bayesian methods using Longer-Job heuristics (16.2).

$R = 100$, $K = 1$, $J = 10$, $S = 10$, and $O = 10$					
Algorithm	f_B	d_B	x_0	x_1	x_2
Delta	6.183	0.133	0.283	0.451	0.266
Taylor 3	6.173	0.083	0.304	0.276	0.420
CPLEX	12.234	0.00	—	—	—

2.4 GMC SOFTWARE EXAMPLE

One applies the C++ optimization system GMC to search for such BHA parameters that reduce the average deviation on a set of randomly generated Flow-Shop problems. The corresponding file $fi.C = fi\_flow.C$ is on the web-site (see section 4.). Figure 16.1 shows how to define data in the file 'gmc/fi_flow.C'.

```
#include <stdio.h> #include <math.h> #include <malloc.h>
#include <string.h> #include "task.h" #define DIM
 3 #define NUMBER_OF_JOBS      7 #define NUMBER_OF_MACHINES
40 #define NUMBER_OF_FCALC     100 #define FALSE
0 #define TRUE               1 int number_of_variables =
DIM;
```

Figure 16.1. Example of the file 'gmc/fi_flow.C.'

2.5 GMJ SOFTWARE EXAMPLE

To optimize BHA parameters for a given Flow-Shop problem, one applies the Java optimization system GMJ2 that reduce the average deviation (Greicius and Luksys, 1999). The operation times are in the file 'jobtimes.txt.' (see Figure 16.2) The local URL address of this file apears opening the GMJ2 task 'FlowShop' (see Figure 16.3).

The optimal proportions of three heuristics: the Monte Carlo, the linear randomization and the "Select-the-Best" are on the output page

```
71.7  62.49 28.65 95.08 51.64 63.97 33.16 36.55 79.05 78.27
67.23 20.99 62.67 4.84  90.02 61.71 24.2  52.85 47.93 99.57
63.54 96.69 33.24 12.29 90.57 57.8  78.79 41.46 46.3  36.32
98.03 22.24 75.09 23.02 22.59 86.13 82.82 16.06 28.36 18.93
76.73 55.52 96.98 43.28 79.6  77.63 99.86 74.95 97.97 20.55
6.68  89.5  5.57  33.93 52.86 29.59 85.1  82.08 25.36 4.96
82.59 82.45 57.4  55.37 75.92 82.6  72.11 77.67 11.6  64.46
94.04 79.15 61.23 86.6  91.56 74.14 56.59 13.99 69.36 45.29
27.89 100.0 77.67 89.27 53.73 47.77 26.92 88.15 74.75 66.31
40.45 97.05 11.04 80.41 25.69 53.58 53.58 31.25 20.52 74.24
22.01 48.93 37.08 11.84 78.14 18.6  15.65 26.69 7.1   32.79
86.37 39.95 71.2  91.72 65.96 87.97 36.91 73.99 85.83 63.97
82.57 98.22 24.91 32.01 64.8  51.18 91.13 62.33 54.05 20.01
7.99  44.9  18.89 4.72  68.69 51.26 49.48 74.39 6.4   71.21
22.39 13.52 75.85 29.9  49.92 57.91 2.81  21.7  26.41 95.1
46.36 95.85 70.24 76.65 47.85 70.24 83.76 59.61 89.51 39.43
97.34 0.15  21.85 48.26 43.36 89.73 85.58 55.82 32.46 80.32
50.56 34.31 93.92 83.44 22.87 20.22 20.37 42.22 90.48 33.84
23.57 9.15  64.97 97.43 77.75 28.3  62.62 56.54 39.98 62.86
83.08 3.45  45.67 36.15 70.0  93.57 2.72  67.69 65.12 42.86
71.17 33.78 90.33 30.31 93.17 76.25 79.7  25.37 61.52 31.51
25.08 27.8  95.49 60.6  3.46  74.63 8.41  98.74 29.05 22.22
98.46 78.16 3.53  65.04 96.55 21.63 49.43 44.92 5.52  8.98
83.61 92.02 90.76 19.81 42.02 40.49 18.65 22.17 87.21 83.77
5.4   54.84 99.76 5.28  14.26 97.87 89.89 80.65 0.46  42.48
82.97 1.61  23.78 11.0  94.76 0.16  55.0  54.76 60.04 74.3
72.16 62.06 42.7  43.16 85.64 68.6  70.22 94.0  4.99  99.76
99.92 54.92 9.68  69.71 44.01 16.17 78.23 20.93 64.09 49.73
```

Figure 16.2. Example of the file 'jobtimes.txt' defining the operation times.

(see Figure 16.4). The convergence line is there, too. These are results of one hundred iterations by the 'Bayes' method.

Figure 16.5 shows how the flow-shop make-span depends on the proportion of the Monte Carlo algorithm. The metod 'Bayes' changes all variables together. Therefore, the projection that depends on one variable is not informative. The 'Projection' menu presents three choices: *Monte Carlo*, *Randomized Heuristics*, and *Greedy Heuristics*.

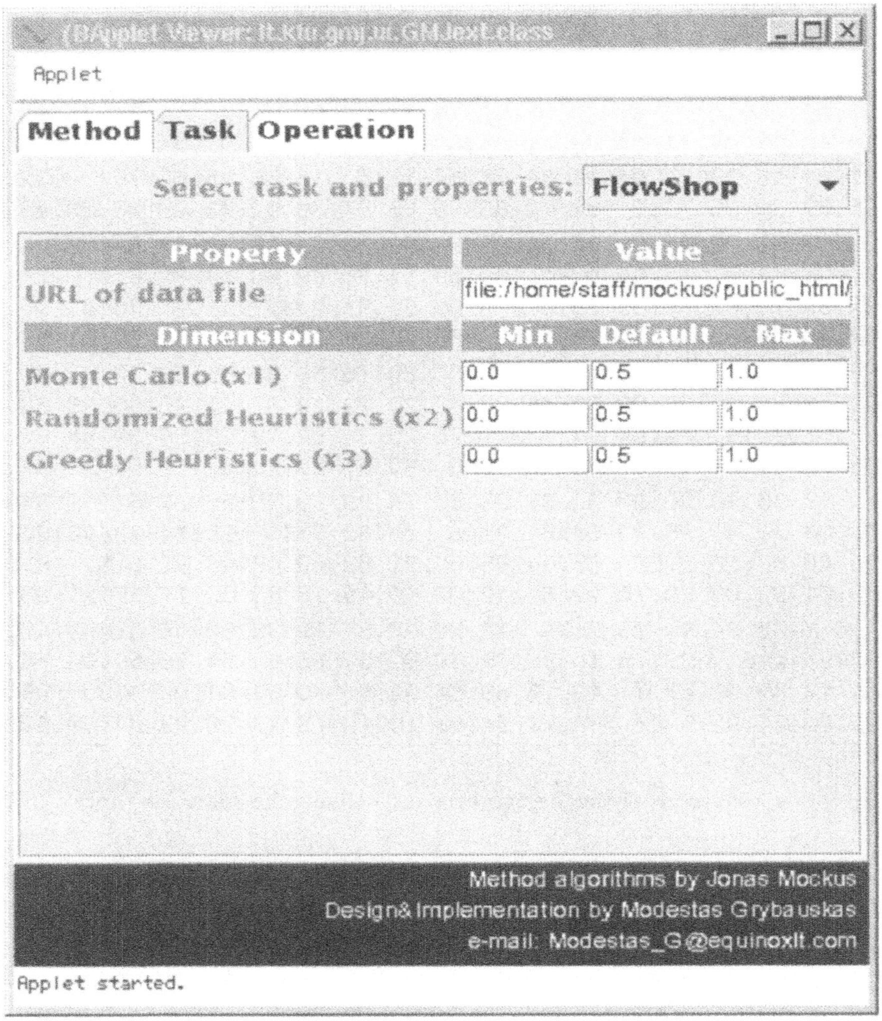

Figure 16.3. Opening the 'FlowShop' task.

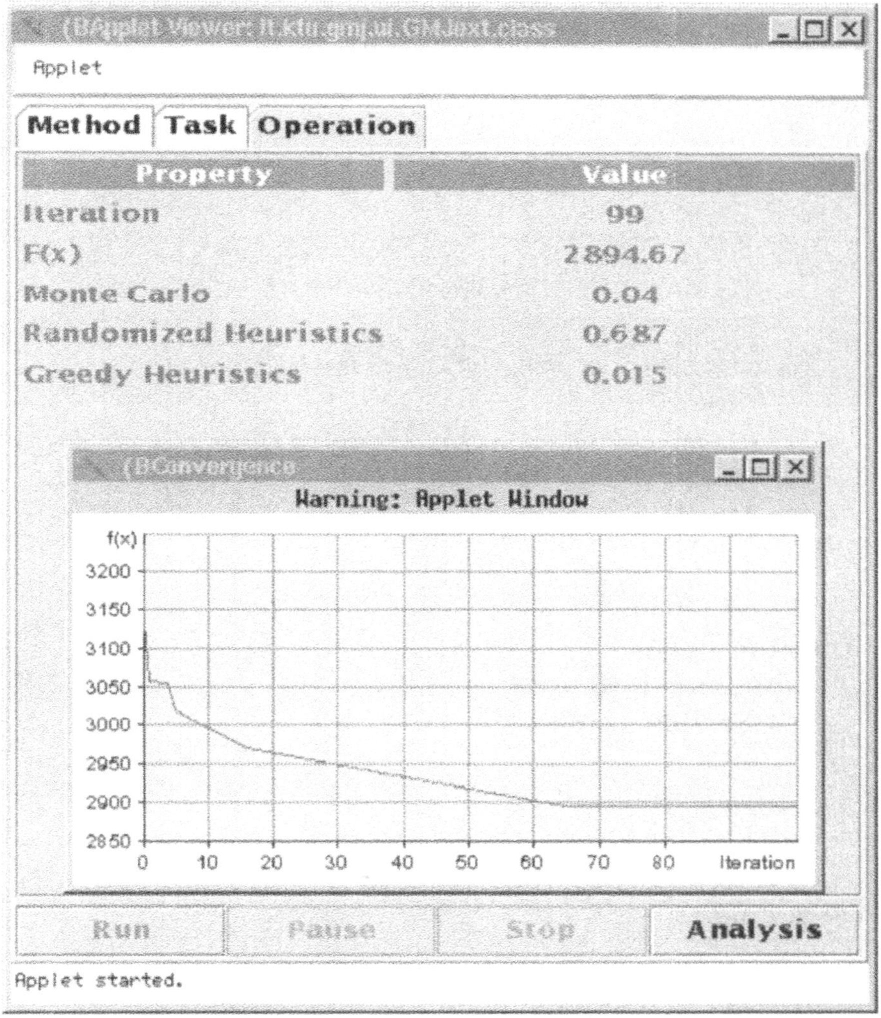

Figure 16.4. Numerical results and convergence line, the 'FlowShop' task.

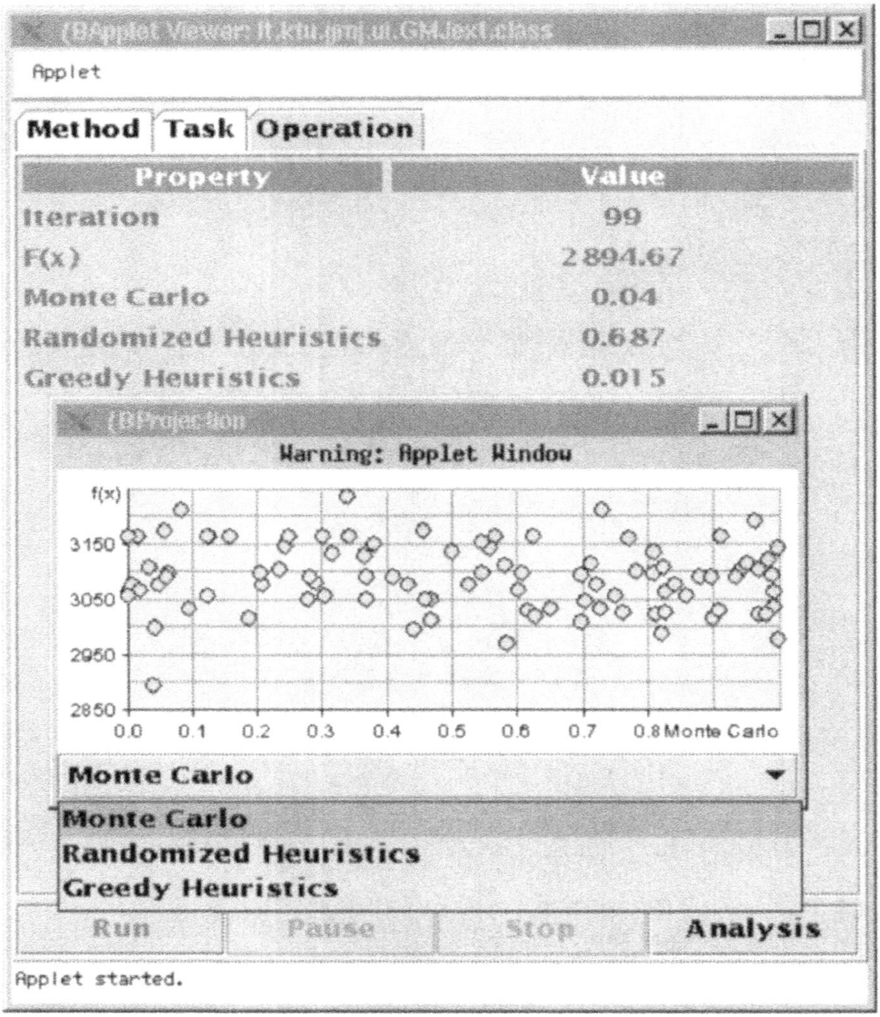

Figure 16.5. Relation of $f(x)$ to the proportiom of Monte Carlo algorithm and the 'Projection' menu.

3. SCHOOL SCHEDULING

School scheduling is the well-known example of a general scheduling problem. Consider the case where the objective is the number of "empty" hours when the teachers wait for the next scheduled lectures. We will call these hours as "teacher windows." We search for such schedules that reduce the sum of teacher windows considering the schedules of fifth to twelfth class of the public high school. Other factors are school-specific and should be included adapting the software to specific schools.

3.1 CONSTRAINTS

There are five constraints:

- no "student windows"are allowed,
- at any given moment a teacher can deliver only one lesson,
- at any given moment a student can take part at only one lesson,
- the "double" lectures (the sequence of two lectures of the same subject) are not allowed (there are some strictly limited exceptions),
- the number of lecture hours per day is limited.

3.2 DATA STRUCTURE

The optimization starts from the existing school schedule. The top picture in Figure 16.11 shows a fragment of the existing weakly schedule of some Lithuanian school.

- the schedule is defined as a string array $Mokytojai[64][38]$,
- the string $Mokytojai[i][1]$ defines the name of the i-th teacher, for example, *L.Dalinkeviciene*, *V.Vilkiene*, e.t.c.
- the string $Mokytojai[i][2]$ defines the subject of the i-th teacher, for example, *lietuviu_k* means Lithuanian, *rusu_k* means Russian, *anglu_k* means English, *matematika* means Mathematics, *informatika* means Informatics, e.t.c.
- the string $Mokytojai[i][3]$ defines the code of the classroom of the i-th teacher, for example, 0 means no special study, A means a study for Informatics, e.t.c.
- the sequence of strings $Mokytojai[i][j]$, $j = 4, ..., 38$ define the class codes of the i-th teacher for all weak, for example, in the class code

$P19c$, the first symbol P means Monday[3], the second symbol 1 defines the first lesson, and the third symbol $9c$ means the class $9c$, teacher windows are denoted by symbols X, symbols Q denote "convenient" lessons, a lesson is called "convenient", if a teacher is free before or after this lesson (see Figures 16.11)[4].

3.3 PERMUTATION AND EVALUATION ALGORITHM

The algorithm follows a general pattern of permutation algorithms related to the Bayesian Heuristic Approach (see, for example, section 2.4)

1. record the current schedule $Mokytojai0$ by reading the data file $Mokytojai[64][38]$,
2. record the current number $f0$ of teacher windows[5],
3. set the initial iteration number $it = 0$,
4. set the current iteration number $it = it + 1,\ it < K$,
5. if $it = K$, stop and print the current schedule,
6. set the initial teacher number $i = -1$,
7. set the current teacher number $i = i + 1$,
8. if $i > 63$ go to the step 4,
9. generate uniformly distributed random number $\xi \in [0, 1]$,
10. go to the step 7, if $\xi < x$,
11. select a window of the teacher i, go to step 7, if there are no windows,
12. select a convenient lesson of the i-th teacher, go to step 7, if there are no convenient lessons,
13. make a permutation by exchanging the window class[6] with the convenient one,
14. test the feasibility conditions listed in section 3.1, go to the step 16, if the permutation is feasible,

[3] The complete list is P for Monday, A for Tuesday, T for Wednesday, K for Thursday, and N for Friday.

[4] In Figures 16.11 there are other symbols, too, such as Y, W, e.t.c.. These figures illustrate an old version of the algorithm (Marcinkevicius, 1999; Dvirmickiene and Zakasauskaite, 1998). Here the new version is described.

[5] A number of windows in the current schedule $Mokytojai0$.

[6] A window class is the class in the teacher window.

15. restore the feasibility by restoring the teacher window and go to the step 7,
16. count the teacher windows in the permuted schedule,
17. compare the number of the teacher windows in the current schedule with the permuted one,
18. replace the current schedule by the permuted one,
    if the current number of teacher windows is less then that in the permuted schedule,
    go to step 7

The algorithm is very simple and natural. It mimics, in a sense, the usual ways to improve existing schedules. There are negative sides, too. There is no proof yet that the algorithm satisfies the convergence conditions (Mockus et al., 1997). To satisfy these conditions the probability to reach any schedule should be positive. Note, that replacing a window class by the convenient one, defined in the step thirteen, a new window for another teacher is opened, as usual.

3.4 SOFTWARE EXAMPLE

The file 'Tvarka.java' (see web-site described in section 4.) implements an algorithm described in (Marcinkevicius, 1999; Dvirmickiene and Zakasauskaite, 1998). This algorithm does similar permutations but is different from the described one. Figure 16.6 shows a fragment of the 'Tvarka.java' file.

One may change the initial data file 'mokytojai.txt' to the new file of the same size without recompiling. The default URL address shows the path of data file in the host server (see Figure 16.6). The URL address can be changed from the screen. One recompiles the *Tvarka.java* file, to change the size of the data file 'mokytojai.txt.'

Users data files can be uploaded by a browser (Venclovas, 1999). Figure 16.7 shows an example using *http://optimum.mii.lt/cgi-bin/fup/fup.cgi?uzd1*. Here the host server *optimum.mii.lt* applies the CGI script to allow users to load their data.

3.5 RUNNING SOFTWARE

The GMJ1 and GMJ2 systems described in Chapter 8) are used for optimization of BHA parameters following these steps:

- select the method and its parameters (see Figure 16.8),

```
public class Tvarka implements Task
{
    private int eMo=64; int sMo=38;
    private int firsttime=0;
    private int iii=0;
    private String url=new String
("http://optimum.mii.lt/~jonas/school4j/mokytojai.txt");
    private String [][] Mokytojai=new String[eMo][sMo];
  private String [][] Mokytojai0=new String[eMo][sMo];
    private int Sdiena;
    private int sdiena;
     private double L0=1000.;
public int K;
 public void customize (PropertyManager manager)
{
manager.add (new IntProperty ("K  number of repetitions",
new TvarkaProvider (this),1,100));
}
```

Figure 16.6. A fragment of the school scheduling program.

- select the task 'Tvarka' and its parameters including the iteration number K, the lower and upper bounds for the randomization parameter x, the default value of x (see Figure 16.9),
- observe the results (see Figure 16.10).

The output page shows the optimal number of teacher windows. To observe the optimal schedule an additional 'analysis object' should be made. That is done clicking the button 'TvarkaAnalyser.' Figures 16.11 show a fragment of school schedules. The top figure is a fragment of the initial schedule[7]. The bottom figure is the corresponding fragment of the schedule optimized using the 'Bayes1" method after five hundred iterations.

[7] In this fragment, the window lessons and the convenient ones are denoted by letters different from those described in Section 3.3, a slightly different algorithm is implemented now.

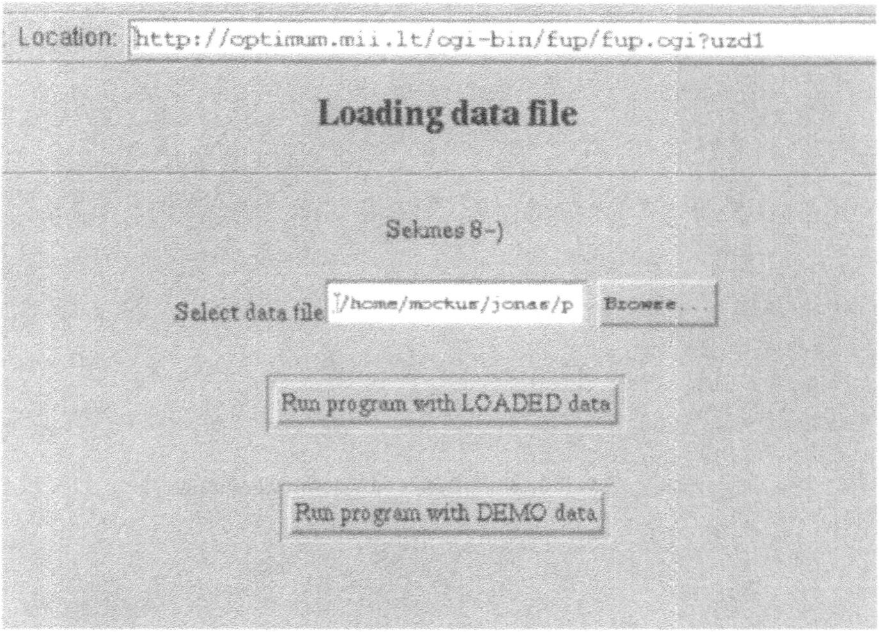

Figure 16.7. Uploading data file.

Figure 16.8. Input window of method selection.

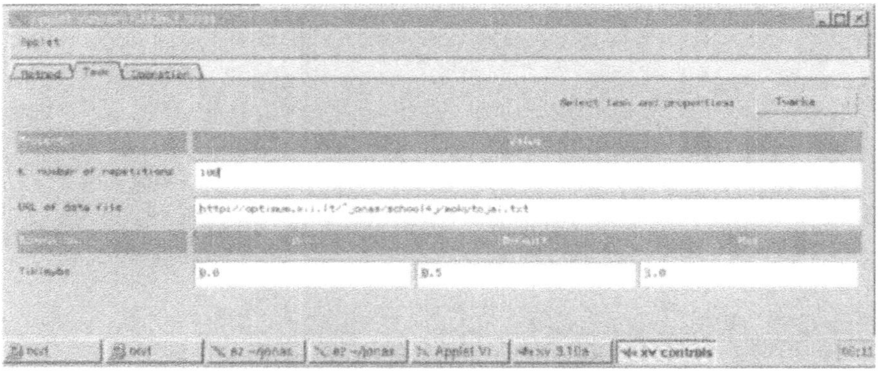

Figure 16.9. Input window of school scheduling.

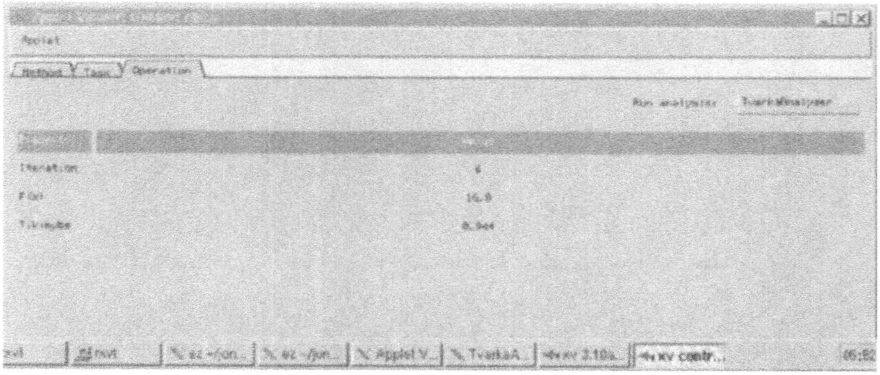

Figure 16.10. Output window of school scheduling.

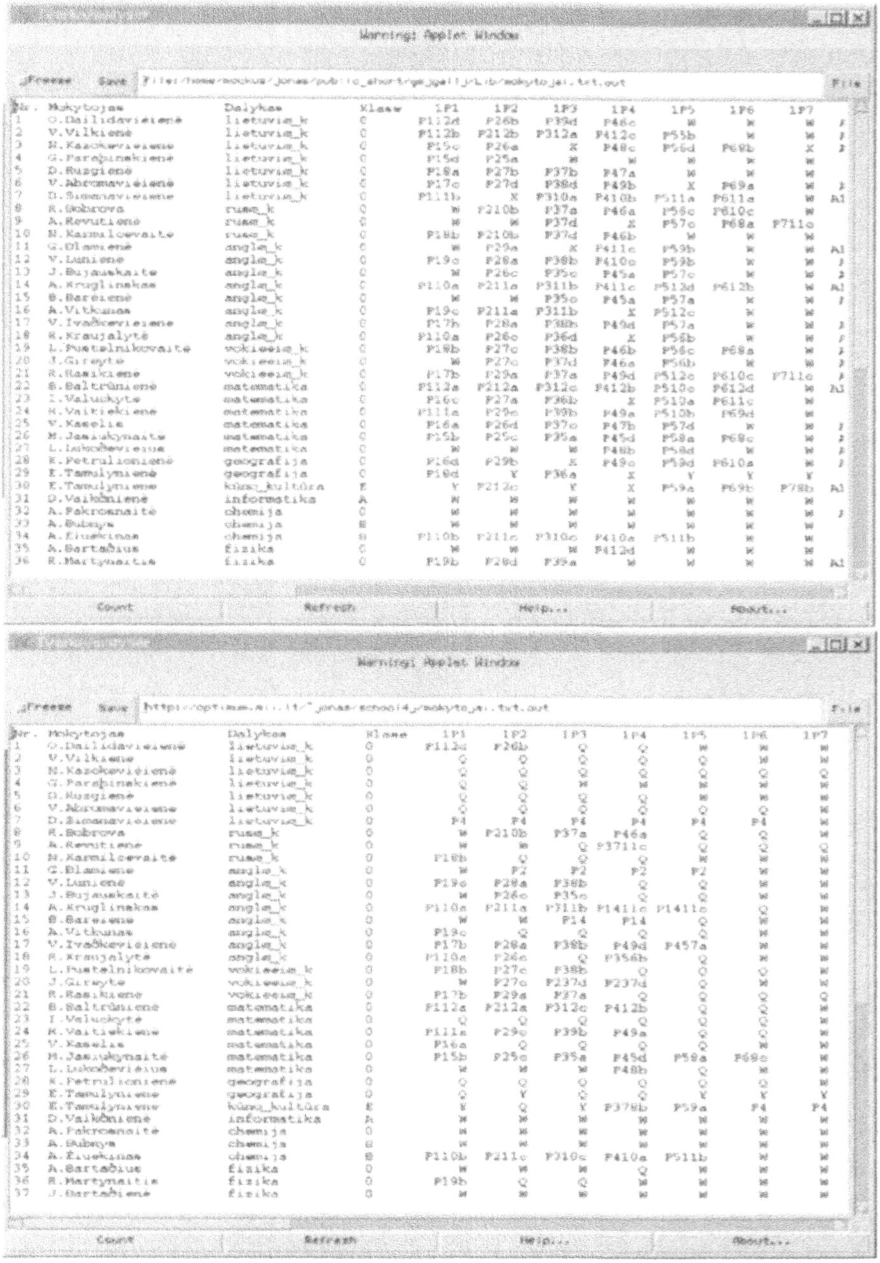

Figure 16.11. **A fragment of school schedules, the initial schedule (the top figure), the optimized schedule (the bottom figure).**

Chapter 17

SEQUENTIAL STATISTICAL DECISIONS MODEL, "BRIDE PROBLEM"

1. INTRODUCTION

Most of the decisions in a personal life and in organizations are made sequentially. That means that, at any given moment, one either makes the final decision, or postpones it hoping for better times. Statistical part of sequential decisions represents uncertainties of future events and a limited reliability of our observations.

The Bride problem is a good example of sequential statistical decisions (Wald, 1947; Wald, 1950). The dynamic programming is a conventional technique to optimize sequential decisions (Bellman, 1957). Applying the dynamic programming to specific problems, one develops specific algorithms, as usual. The algorithms, similar to these of bride's decision making, can be applied choosing the optimal time to buy durable goods, such as cars, houses, computers, to start new business, e.t.c..

Typical industrial applications of sequential decisions are the optimal stopping rules. For example, one can consider stopping of some nuclear plant as a marriage. One can evaluate stopping cost directly. Risks involved in running the plant, when one observes some deviations from the normal routine, can be considered as risks of waiting for the next proposal. That means that one hopes for something better while risking a bad outcome.

2. AVERAGE UTILITY

The Bride problem is to maximize the average utility of marriage by the optimal choice of a groom. Denote the actual goodness of a groom i by ω_i. Denote by s_i the bride's impression about the groom i. $p(\omega_i$ is a prior probability density of goodness ω_i. $p_s(s_i|\omega_i)$ is a probability density of impression s_i. Assume that goodness of different grooms are independent and identically distributed. This means that a prior probability density of goodness is

$$p(\omega_i, \omega_j) = p(\omega_i)p(\omega_j),\qquad (17.1)$$
$$p(\omega_i) = p(\omega_j) = p(\omega).$$

Suppose that impressions about the goodness of grooms are independent and identically distributed. Then the probability density of an impression s_i, given the goodness ω_i, is

$$p_s(s_i, s_j|\omega) = p_s(s_i|\omega)p_s(s_j|\omega),\qquad (17.2)$$
$$p_s(s_i|\omega) = p_s(s_j|\omega) = p(s|\omega).$$

Assume the Gaussian prior probability density of goodness

$$p(\omega) = \frac{1}{\sqrt{2\pi}\sigma_0} e^{-1/2(\frac{\omega-\alpha_0}{\sigma_0})^2}.\qquad (17.3)$$

Here σ_0 is a prior variance and α_0 is a prior mean of goodness. For example, $\alpha_0 > 0$ shows an optimistic bride. $\alpha_0 < 0$ shows that a bride is pessimistic. Suppose, that a prior probability density of bride impressions is

$$p(s|\omega) = \frac{1}{\sqrt{2\pi}\sigma} e^{-1/2(\frac{s-\omega}{\sigma})^2}.\qquad (17.4)$$

Here σ^2 is a variance of impressions around the true goodness ω.

Assume that both the groom goodness and the bride impression are random variables depending on many independent factors. This explains the Gaussian distributions (17.3) and (17.4). One defines a posterior probability density of goodness ω, given the impression s, by the Bayesian formula (Bayes, 1783)

$$p(\omega|s) = \frac{p(s|\omega)p(\omega)}{p_s(s)}.\qquad (17.5)$$

Here

$$p_s(s) = \int_{-\infty}^{\infty} p(s|\omega)p(\omega)d\omega.\qquad (17.6)$$

3. SINGLE-MARRIAGE CASE

Denote by d_i the bride's decision about the groom i

$$d_i = \begin{cases} 1, & \text{if } \textit{bride marry the groom } i, \\ 0, & \text{otherwise.} \end{cases} \tag{17.7}$$

Suppose that

$$\sum_{i=1}^{m} d_i = 1. \tag{17.8}$$

The last condition means that brides marry, and marry only once. Formally condition (17.8) defines the set of feasible decisions when the nth groom proposes

$$D_{N-n} = \begin{cases} 0 \textit{ and } 1, & \text{if } g_{N-n} = 0 \\ 0, & \text{if } g_{N-n} = 1, \end{cases} \tag{17.9}$$

Here g_{N-n} is the marriage index

$$g_{N-n} = 1 - \sum_{i=1}^{N-n-1} \tag{17.10}$$

The marriage index is zero, if the bride is marred. This prevents repeated marriages.

3.1 BELLMAN'S EQUATIONS

The expected utility function is $u(s)$. Here s is the impression made by the successful groom[1].

$$u(s) = \int_{-\infty}^{\infty} \omega p(\omega|s) d\omega. \tag{17.11}$$

Denote by $u_N(s)$ the expected utility function, if the impression of the last groom is s

$$u_N(s) = \int_{-\infty}^{\infty} \omega p(\omega|s) d\omega. \tag{17.12}$$

Comparing (17.11) and (17.12) one observes that

$$u(s) = u_N(s). \tag{17.13}$$

[1] By the "successful groom" we mean the groom that a bride marries.

This follows from independence assumptions (17.1) and (17.2). Denote by u_{N-1} the expected utility, if the impression of the $(N-1)$th groom is s and a bride is making the optimal decision $d = d_{N-1}(s) \in D_{N-1}$

$$u_{N-1}(s) = \max_d (du(s) + (1-d)u_N), \tag{17.14}$$

$$d_{N-1}(s) = arg \max_d (du(s) + (1-d)u_N). \tag{17.15}$$

Here

$$u_N = \int_{-\infty}^{\infty} u_N(s) p_s(s) ds. \tag{17.16}$$

Following the same pattern, we define the expected utility, if the impression of the $(N-n)$th groom is s and the bride is making the optimal decision $d = d_{N-n}(s) \in D_{N-n}$

$$u_{N-n}(s) = \max_d (du(s) + (1-d)u_{N-n+1}), \tag{17.17}$$

$$d_{N-n}(s) = arg \max_d (du(s) + (1-d)u_{N-n+1}). \tag{17.18}$$

Here

$$u_{N-n+1} = \int_{-\infty}^{\infty} u_{N-n+1}(s) p_s(s) ds. \tag{17.19}$$

Note, that the utility $u(s)$ of accepting a proposal in expressions (17.17) and (17.18) is a function only of impression s. It does not depend on the proposal number $N-n$. That follows from independence assumptions (17.1) and (17.2). Solving these recurrent equations one defines the sequence of optimal decision functions $d_{N-n}(s) \in D_{N-n}$ and the expected utilities $u_{N-n}(s)$. This should be done for all possible impressions $s \in (-\infty, \infty)$ and for all numbers $n = 1, ..., N-1$. One cannot do that in continuous case. Therefore, one uses a discrete approximation

3.2 DISCRETE APPROXIMATION

From expressions (17.11) and (17.12), replacing the integrals by sums one obtains that

$$u_N(s) = u(s) = 2M/K \sum_{k=1}^{K} \omega_k p(\omega_k|s). \tag{17.20}$$

From expression (17.16)

$$u_N = 2M/K \sum_{k=1}^{K} u_N(s_k) p_s(s_k). \tag{17.21}$$

From expression (17.19)

$$u_{N-n+1} = 2M/K \sum_{k=1}^{K} u_{N-n+1}(s_k)p_s(s_k). \tag{17.22}$$

Here $\omega_k \in [-M, M]$, $\omega_1 = -M$, $\omega_K = M$ and $s_k \in [-M, M]$, $s_1 = -M$, $s_K = M$. That is a discrete approximation of the recurrent equations. All possible impressions $s_k \in [-M, M]$ and all numbers $n = 1, ..., N-1$ are considered. One sets the number of iterations K by the accuracy needed.

The results are sequences of optimal decision functions $d_{N-n}(s_k)$ and the expected utilities $u_{N-n}(s_k)$. These sequences are stored in a set of arrays which define how the optimal decisions d and the expected utilities u depend on the possible impressions s_k, $k = 1, ..., K$. Doing this, one avoids the repeated calculations. Large arrays is a disadvantage of this procedure.

3.3 INCLUDING THE WAITING COST

The waiting losses are important in many real-life sequential decision problems. Denote by c the loss of waiting for the next groom. Including this parameter into Bellman equations one obtains

$$u_{N-1}(s) = \max_d (du_N(s) + (1-d)(u_N - c)), \tag{17.23}$$

$$d_{N-1}(s) = arg\max_d (du_N(s) + (1-d)(u_N - c)). \tag{17.24}$$

In a similar way one defines the expected utility if the impression of the $(N-n)$-th groom is s and the bride is making the optimal decision $d_{N-n}(s)$

$$u_{N-n}(s) = \max_d (du_N(s) + (1-d)(u_{N-n+1} - c)), \tag{17.25}$$

$$d_{N-n}(s) = arg\max_d (du_N(s) + (1-d)(u_{N-n+1} - c)). \tag{17.26}$$

The other expressions remains the same.

3.4 NON-LINEAR CASE

Expression (17.12) was defined assuming the linear bride's utility function. It was supposed that bride's utility is equal to the goodness of groom $u(\omega) = \omega$. In the real life, utility functions are nonlinear (see chapter 13) and non-convex, as usual. Then expression (17.12) is replaced by this integral

$$u_N(s) = \int_{-\infty}^{\infty} u(\omega)p(\omega|s)d\omega. \tag{17.27}$$

4. MULTI-MARRIAGE CASE

Condition (17.8) implies that brides marry and marry only once. No divorce is permitted. That is wrong, if the "marriage" represents baying a Personal Computer (PC). We illustrate this by a simple "Buy-a-PC" example.

4.1 " BUY-A-PC" EXAMPLE

Define PC parameters by a vector $g = (g_1, g_2, g_3)$. Here g_1 is the speed of CPU in MHz, g_2 is the volume of RAM in MB, and g_3 is the volume of HD in GB. Express a subjective utility of PC by the weighted sum $\omega = a_1 g_1 + a_2 g_2 + a_3 g_3$. Here a_i are expressed in \$ per unit.

Assume that a prior probability density of PC utilities is Gaussian

$$p_t(\omega) = \frac{1}{\sqrt{2\pi}\sigma_0} e^{-1/2(\frac{\omega-\alpha_t}{\sigma_0})^2}. \tag{17.28}$$

Here σ_0 is a prior variance and α_t is a prior mean of PC utilities. Suppose that $\sigma_0 = constant$ and that $\alpha_t = \alpha_0 + \alpha_1 t$. This means that expected PC utility is increasing linearly. The expected diversity remains the same.

The PC price in \$ is denoted by l. Suppose that the price of PC depends linearly on the weighted sum

$$l = b_1 g_1 + b_2 g_2 + b_3 g_3. \tag{17.29}$$

Here parameters b_i are defined in \$ per unit and reflect the market prices of CPU, RAM, and HD. Expressing the price l as a function of utility ω:

$$l = h_0 \omega. \tag{17.30}$$

Here

$$h_0 = \frac{b_1 g_1 + b_2 g_2 + b_3 g_3}{a_1 g_1 + a_2 g_2 + a_3 g_3}. \tag{17.31}$$

Assume that one observes the utility ω exactly. This means that the impression $s = \omega$ and that impression errors $\sigma = 0$ in expression (17.2). That simplifies the problem.

4.2 BELLMAN'S EQUATIONS

The expected utility function is $u(\omega, q)$. Here ω is the utility of a new PC. q is the utility of the old PC, to be replaced by the new one. Consider a "horizon" of N years. During a year one can change PC only once, if one wishes.

Denote by $u_N(\omega, q)$ the maximal expected utility in the year N

$$u_N(\omega, q) = \max_d (d\omega + (1-d)u_{N+1}). \tag{17.32}$$

There are two possible decisions $d = \{0, 1\}$. The decision $d = 1$ means to bay a new PC. The utility of the new PC is ω. The utility of the old one is q. The utility of the decision $d = 0$ to keep the old PC is $u = q + l$. Here l is the price of the new PC defined by (17.30). It is assumed that we abandon the old PC as soon as we obtain the new one. Therefore, one "wins" the price l of the new PC by using the old PC. The optimal decision

$$d_N(r) = \begin{cases} 1, & \text{if } r \geq 0, \\ 0, & \text{if } r < 0. \end{cases} \tag{17.33}$$

Here $\omega = \omega_N$ is the utility of the new PC available in the year N. $u_{N+1} = q_N + l_N$ is the utility of keeping the old PC in the year N.

$$r = \omega_N - u_{N+1} \tag{17.34}$$

is the difference between the utility ω_N of baying new PC and the utility $u_{N+1} = q_N + l_N$ of keeping the old one.

Denote by $u_{N-1}(\omega, q)$ the maximal expected utility in the year $N-1$.

$$u_{N-1}(\omega, q) = \max_d (d\omega + (1-d)u_N(q). \tag{17.35}$$

Here $u_N(q)$ is the maximal expected utility in the year N, if the utility of the old PC is q.

$$u_N(q) = E\, u_N(\omega, q) = \int_{-\infty}^{\infty} u_N(\omega, q) p_N(\omega) d\omega. \tag{17.36}$$

$p_N(\omega)$ is a prior probability density of ω defined by expression (17.28) at a time $t = N$.

$$q = \begin{cases} \omega_{N-1}, & \text{if } r \geq 0, \\ q_{N-1}, & \text{if } r < 0. \end{cases} \tag{17.37}$$

$r = \omega_{N-1} - u_N(q)$ is the difference between the utility ω_{N-1} of baying and the expected utility $u_N(q)$ of keeping .

The parameter r depends on two variables ω and q. Therefore, the "One-Step-Ahead" approximation is suggested where $r = \omega_{N-1} - q_{N-1} - l_{N-1}$. Here one assumes that the next step is the last one. Thus, r is expressed by the last step formula (17.34). The approximation error depends on many factors. One estimates it by experimentation.

The optimal decision

$$d_{N-1}(r) = \begin{cases} 1, & \text{if } r \geq 0, \\ 0, & \text{if } r < 0. \end{cases} \tag{17.38}$$

Following the same pattern, one defines the maximal expected utility in $(N - n)$th year

$$u_{N-n}(\omega, q) = \max_d (d\omega_{N-n} + (1 - d)u_{N-n+1}(q)), \tag{17.39}$$

where

$$u_{N-n+1}(q) = E\ u_{N-n+1}(\omega, q) = \int_{-\infty}^{\infty} u_{N-n+1}(\omega, q) p_{N-n+1}(\omega) d\omega. \tag{17.40}$$

Here the utility of new PC is ω, and the utility of the old one is q

$$q = \begin{cases} \omega_{N-1}, & \text{if } r \geq 0, \\ q_{N-1}, & \text{if } r < 0. \end{cases} \tag{17.41}$$

The optimal decision

$$d_{N-n}(r) = \begin{cases} 1, & \text{if } r \geq 0, \\ 0, & \text{if } r < 0. \end{cases} \tag{17.42}$$

The optimal decision $d_n(r)$ is a function of difference $r = \omega_{N-n} - u_{N-n+1}(q)$ that depends on two parameters ω and q. Therefore, one defines it as a function of critical pairs (ω, q) depending on time $t = n$.

Solving these recurrent equations one defines the sequence of optimal decision functions $d_{N-n}(r)$ and the expected utilities $u_{N-n}(\omega, q)$. This should be done for all possible values of ω and q.

One applies the discrete approximation such as in Section 3.2. The optimal decision functions depend on two variables ω and q. Therefore, one needs K times more calculations to obtain the same accuracy as in the single marriage case (see Section 3.2).

The "One-Step-Ahead" approximation, assuming that $r = \omega_{N-n} - q_{N-n} - l_{N-n}$, reduces the computing time to a level of single marriage case. However, one needs to estimate the approximation error.

To solve real life problems the "directly unobservable" factors should be included into the "Buy-a-PC" model. For example , one can estimate the expected life-span $T(s)$ of PC, given the impression s, by the Bayesian formula

$$T = \int_{-\infty}^{\infty} \tau p(\tau|s) d\tau, \tag{17.43}$$

$$p(\tau|s) = \frac{p(s|\tau)p(\tau)}{p_s(s)}. \tag{17.44}$$

Here τ is the life-span of PC and s is the user's impression about its reliability. The impression s depends on the warranty, the reputation of the manufacturer and on some subjective factors, too.

5. SOFTWARE EXAMPLES

Two interactive versions of the Bride problem are implemented as Java applets. These, and the corresponding $C++$ software are on websites (see section 4.).
Figure 17.1 shows the input window of the Bride problem (Kajokas, 1997)
Figure 17.2 shows simulated results of the "optimal marriage".
Figure 17.3 shows the optimal decision function defining how the critical "goodness" of a groom depends on his number.
Figure 17.4 shows the help window of the "Buy-a-Car" version of the Bride problem (Mikenas, 1998).
Figure 17.5 shows the input window of the "Buy-a-Car" problem.
Figure 17.6 shows the output window of the "Buy-a-Car" problem.
Figure 17.7 shows the optimal decision function of the "Buy-a-Car" problem.

The "Buy-a-Car" algorithm is just a first approximation. It should be improved. For example, improving the By-a-PC model one should

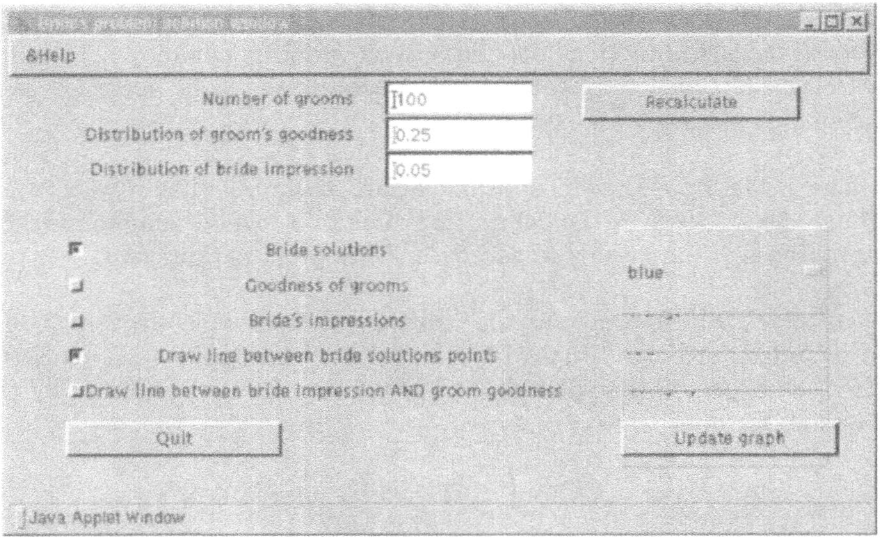

Figure 17.1. The input window of the Bride problem.

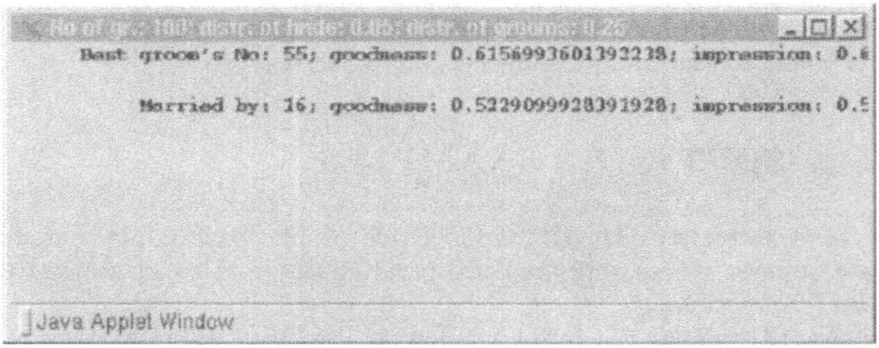

Figure 17.2. The result of statistical simulation of the Bride problem.

- remove the "single marriage" condition (17.8),
- add "Change-a-Car" cost,
- express desirable car properties (see Figure 17.5), using conditions of Pareto optimality (Mockus, 1989b),
- extend the property list by including properties that are not directly observable, such as the reliability of a car,
- keep the "No-Return" rule.

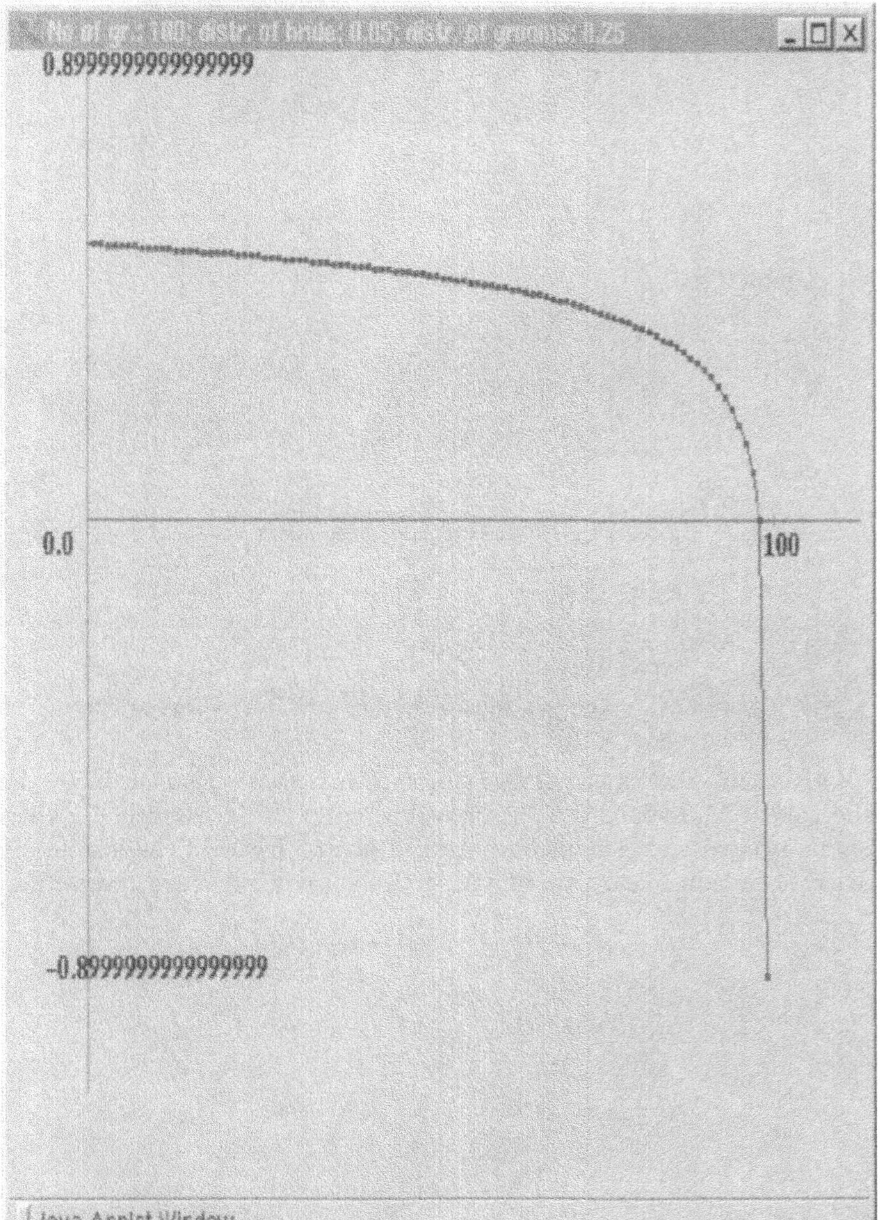

Figure 17.3. The optimal decision function of the Bride problem.

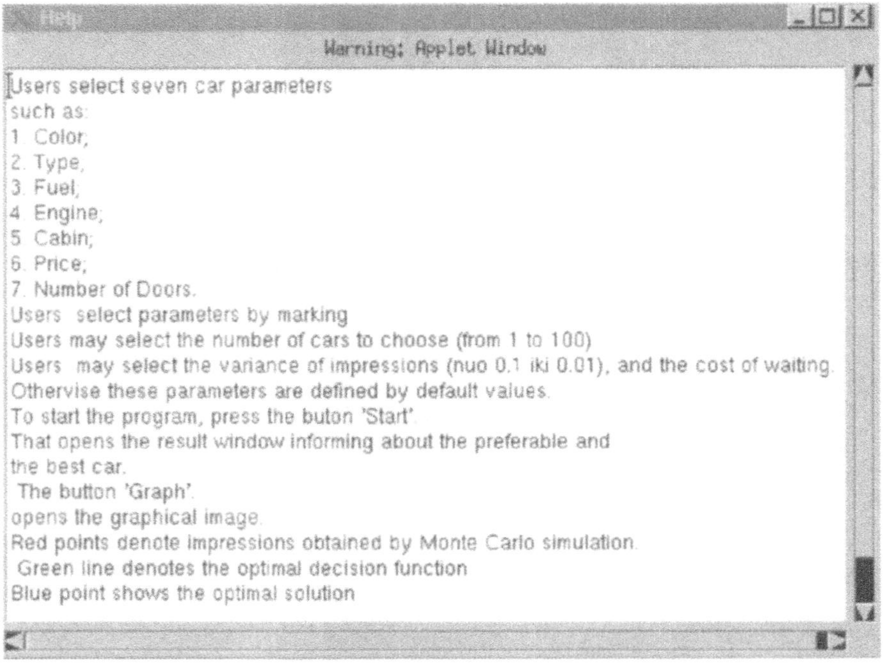

Figure 17.4. The help window solving the "Buy-a-Car" problem.

"No-Return" means that one cannot return to a situation which he or she rejected previously. The "No-Return" rule is essential in the dynamic programming. One considers "Return" cases in the framework of discrete or non-linear programming. That requires more computing, as usual.

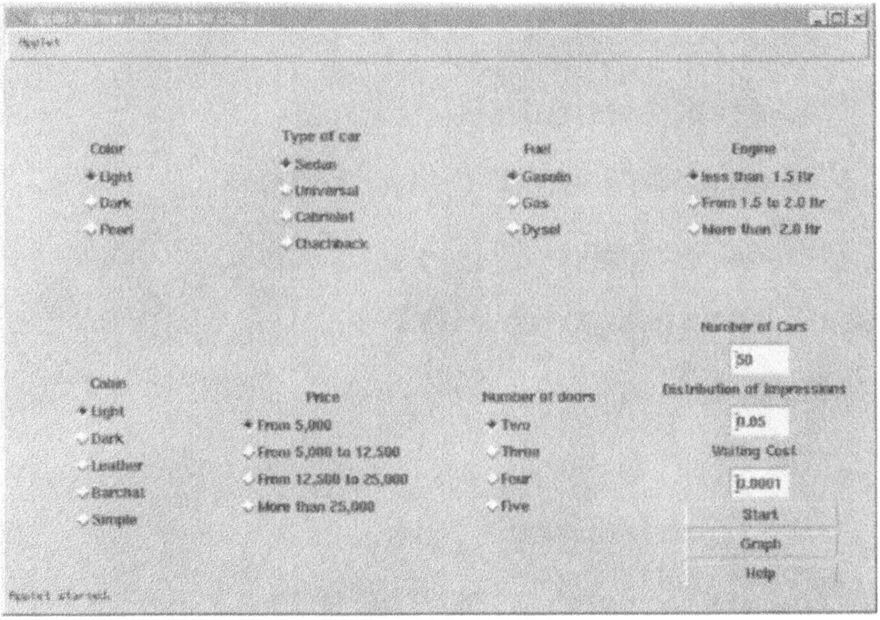

Figure 17.5. The input window solving the "Buy-a-Car" problem.

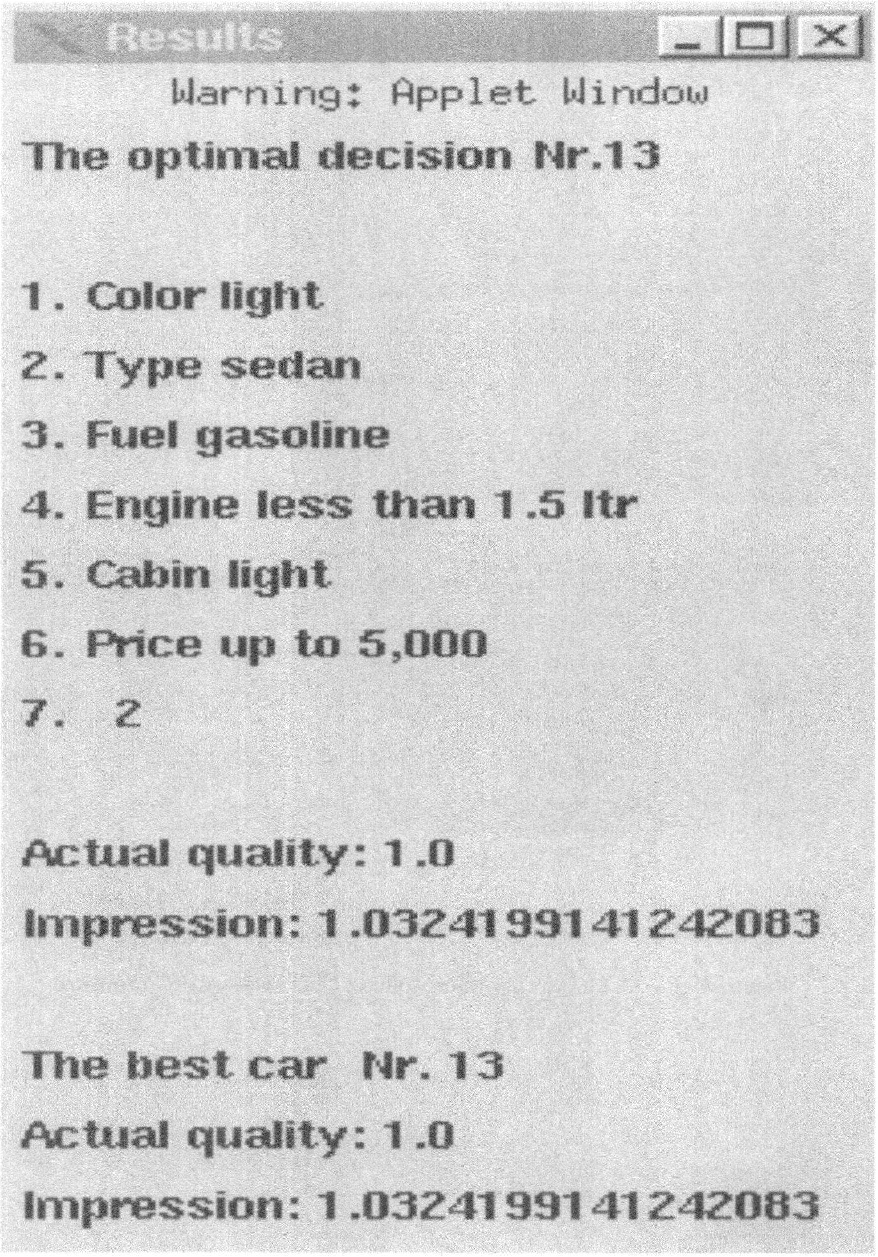

Figure 17.6. The output window of the "Buy-a-Car" problem.

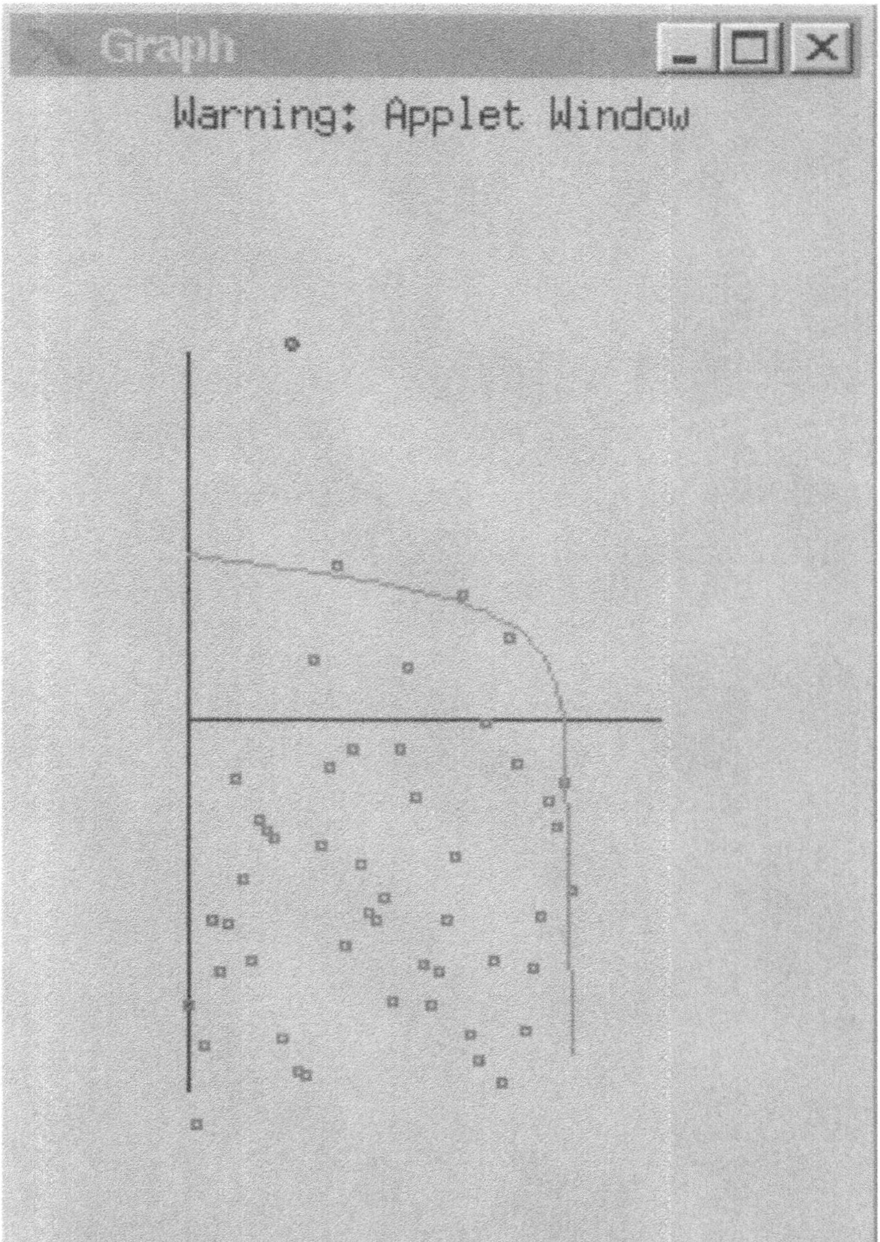

Figure 17.7. The optimal decision function of the "Buy-a-Car" problem.

References

Adamiecki, K. (1931). Harmonygraph. *Przeglad Organizaciji.*

Androulakis, I. and Venkatasubramanian, V. (1991). A genetic algorithm: Framework for process design and optimization. *Computers in Chemical Engineering*, 15:217–228.

Baker, K. R. (1974). *Introduction to Sequencing and Scheduling.* John Wiley & Sons, New York.

Bayes, T. (1783). An essay towards solving a problem in the doctrine of chances. *Phil. Transactions of Royal Society*, 53:370–418.

Bellman, R. (1957). *Dynamic Programming.* Princeton University Press, Princeton, New Jersey.

Biggs, M. C. (1975). Constrained minimisation using recursive quadratic programming: some alternative subproblem formulations. In *Towards global optimisation.* North Holland, Amsterdam.

Brouwer, L. (1912). Uber abbildung von mannigfaltigkeiten. *Math. Ann.*, 71:97–115.

Browder, F. (1968). The fixed point theory of multivalued mappings in topological vector spaces. *Math. Ann.*, 177:283–302.

Cheung, Y.-W. (1993). Long memory in foreign exchange rates. *Journal of Business and Economic Statistics*, 1:93–101.

Cheung, Y.-W. and Lai, K. (1993). Fractional co-integration analysis of purchasing power parity. *Journal of Business and Economic Statistics*, 1:103–112.

DeGroot, M. (1970). *Optimal Statistical Decisions.* McGraw-Hill, New York.

Diaconis, P. (1988). Bayesian numerical analysis. In *Statistical Decision Theory and Related Topics*, pages 163–175. Springer Verlag.

Diebold, F. X. and Rudebusch, G. D. (1989). Long memory and persistence in aggregate output. *Journal of Monetary Economics*, 24:189–209.

Dixon, L. and Szego, G. (1978). *Towards global optimisation 2.* North Holland, Amsterdam.

Dvirmickiene, D. and Zakasauskaite, A. (1998). School scheduling problem. Technical report, Vytautas Magnus University, Faculty of Informatics, Lithuania, Kaunas. (in Lithuanian).

Dzemyda, G. and Senkiene, E. (1990). Simulated annealing for parameter grouping. In *Transactions. Information Theory Statistical Decision Theory Random Processes*, pages 373–383, Praque.

Dzemyda, G., Vaitiekunas, F., Valincius, G., Vysniauskas, J., Juzefovic, D., Kurcerenko, V., Kuziakin, O., and Filatov, N. (1984). Solution of problems of optimal design and selection of model parameters using the package of applied programs minimum. *Optimal Decision Theory, Institute of Mathematics and Cybernetics, Vilnius, Lithuania*, 10:77–98. (in Russian).

Ermoljev, Y. and Wets, R.-B. (1988). *Numerical Techniques for Stochastic Optimization.* Springer-Verlag, Berlin- New York-London.

Fishburn, P. (1964). *Decision and Value Theory.* Wiley, New York.

Floudas, C., Pardalos, P., Adjiman, C., Esposito, W., Z.H. Gu, u., Harding, S., Klepeis, J., Meyer, C., and Schweiger, C. (1999). *Handbook of test problems in local and global optimization.* Kluwer Academic Publishers, Dordrecht-Boston-London.

Forgo, F., Szep, J., and Szidarovszky, F. (1999). *Introduction to the Theory of Games*. Kluwer Academic Publishers.

Fox, R. and Taqqu, M. (1986). Large-sample properties of parameter estimates for strongly dependent stationary gaussian time series. *Annals of Statistics*, 14:517–532.

Friedman, A. (1971). *Dfferential Games*. Wiley-Interscience, New York.

Galperin, E. and Zheng, Q. (1991). *Global Solutions in Optimal Control and Games*. Presses de l'Universite du Quebec a Montreal, Montreal.

Geweke, J. and Porter-Hudak, S. (1983). The estimation and application of long memory time series models. *Journal of Time Series Analysis*, 4:221–238.

Glover, F. (1994). Tabu search: improved solution alternatives. In *Mathematical Programming. State of the Art 1994*, pages 64–92. University of Michigan.

Gnedenko, B. and Kovalenko, I. (1987). *Introduction to the Theory of Queuing*. Nauka, Moscow. (in Russian).

Goldberg, D. E. (1989). *Genetic Algorithms in Search, Optimization, and Machine Learning*. Addison-Wesley, Reading, MA.

Greicius, A. and Luksys, R. (1999). Conversion from c++ to java of the flow-shop model. Technical report, Vytautas Magnus University, Faculty of Informatics, Kaunas, Lithuania.

Grybauskas, M. (1998). Global minimizer for java (gmj) version 1.1. Technical report, Institute of Mathematics and Informatics, Akademijos 5, Vilnius, Lithuania.

Hajdu, M. (1997). *Network Scheduling Techniques for Construction Project Management*. Kluwer Academic Publishers.

Hajek, O. (1975). *Pursuit Games*. Academic Press, New York.

Helman, P., Moret, B., and Shapiro, H. (1993). An exact characterization of greedy structures. *SIAM Journal of Discrete Mathematics*, 6:274–283.

Herings, P. (1994). A gobally and universally stable price ajustment process. Technical Report 9452, Center for Economic Research, Tilburg University, The Netherlands, Tilburg.

Himmelblau (1972). *Applied Nonlinear Programming*. McGraw-Hill.

Holand, J. (1975). *Adaptation in Natural and Artificial Systems*. University of Michigan Press, An Arbor, MI.

Horst, R., Pardalos, P. M., and Thoai, N. V. (1995). *Introduction to Global Optimization*. Kluwer Academic Publishers, Dordrecht.

Hosking, J. (1981). Fractional differencing. *Biometrika*, 68:165–176.

Hosking, J. (1984). Modeling persistence in hydrological time series using fractional differencing. *Water Resources Research*, 20:1898–1908.

Isaacs, R. (1965). *Dfferential Games*. Wiley, New York.

Janacek, G. (1982). Determining the degree of differencing for time series via the long spectrum. *Journal of Time Series Analysis*, 3:177–188.

Kajokas, V. (1997). Java1.1 software for the bride problem. Technical report, Kaunas Technological University, Faculty of Informatics, Kaunas, Lithuania. (in Lithuanian).

Kakutani, S. (1941). A generalization of brouwer's fixed point theorem. *Duke Journal of Mathematics*, 8:457–459.

Ko, K.-I. (1991). *Complexity Theory of Real Functions*. Birkhauser, Boston.

Koop, G., Ley, E., Osiewalski, J., and Steel, M. F. (1994). Bayesian analysis of long memory and persistence using arfima models. Technical report, Department of Economics, University of Toronto.

Krapauskas, M. (1997). Java1.0 software for global optimization. Technical report, Kaunas Technological University, Faculty of Informatics, Kaunas, Lithuania. (in Lithuanian).

Krasovski, N. (1970). *Game Problems on the encounter motions*. Nauka, Moscow. (in Russian).

Kushner, H. (1964a). A new method of locating the maximum point of an arbitrary multi-peak curve in the presence of noise. *J. of Basic Engineering*, 86:97–100.

Kushner, H. (1964b). A versatile stochastic model of a function of unknown and varying form. *J. of Mathematical Analysis and Applications*, 5:150–167.

Kuzminaite, V. (1999). Conversion from c++ to java of the algorithm of the arma model. Technical report, Vytautas Magnus University, Faculty of Informatics, Kaunas, Lithuania.

Li, W. and McLeod, A. I. (1986). Fractional time series modeling. *Biometrica*, 73:217–221.

Liu, J. (1989). On the existence of general multiple bilinear time series. *Journal of Time Series Analysis*, 10:341–355.

Malisauskas, V. (1998). Software for the knapsack problem. Technical report, Kaunas Technological University, Faculty of Informatics, Kaunas, Lithuania. (in Lithuanian).

Marcinkevicius, V. (1999). Improving the school scheduling problem. Technical report, Kaunas Technological University, Faculty of Informatics, Kaunas, Lithuania. (in Lithuanian).

Margelevicius, M. (1999). Conversion from c++ to java of the algorithm of nonlinear programming. Technical report, Kaunas Technological University, Faculty of Informatics, Kaunas, Lithuania.

Matulevicius, R. (1999). Differential game model. Technical report, Vytautas Magnus University, Faculty of Informatics, Kaunas, Lithuania.

Mavridou, T., Pardalos, P., Pitsoulis, L., and Resende, M. (1998). A GRASP for the biquadratic assignment problem. *European Journal of Operations Research*, 105:613–621.

Michael, J. (1976). *The Computation of Fixed Points and Applications.* Springer-Verlag, Berlin.

Mikenas, M. (1998). Java1.1 software for the bay-a-car version of the bride problem. Technical report, Kaunas Technological University, Faculty of Informatics, Kaunas, Lithuania. (in Lithuanian).

Mockus, A. and Mockus, L. (1990). Design of software for global optimization. *Informatica*, 1:71–88.

Mockus, J. (1967). *Multi-modal Problems in Engineering Design.* Nauka, Moscow. (in Russian).

Mockus, J. (1989a). *Bayesian approach to global optimization.* Kluwer Academic Publishers, Dordrecht-London-Boston.

Mockus, J. (1989b). Bayesian approach to global optimization and application to constrained and multi-objective problems. In *Abstracts, the Fifth International Conference on Stochastic Programming in Ann Arbor, August 13-18, 1989.*

Mockus, J. (1997). A set of examples of global and discrete optimization: application of bayesian heuristic approach ii. *Informatica*, 8(4):495–526.

Mockus, J., Eddy, W., Mockus, A., Mockus, L., and Reklaitis, G. (1997). *Bayesian Heuristic Approach to Discrete and Global Optimization.* Kluwer Academic Publishers, Dordrecht-London-Boston.

Mockus, J. and Soofi, A. (1995). The long-run economic relationships: An optimization approach to fractional integrated and bilinear time series. *INFORMATICA*, 6:61–70.

Nash, J. (1950). Equilibrium points in n-person games. *Proc. Nat. Acad. Sci. USA*, 36:48–49.

Nash, J. (1951). Noncooperative games. *Annals of Mathematics*, 54:286–295.

Nemajunas, I. and Remeikis, N. (1999). Conversion from c++ to java of the school scheduling model. Technical report, Vytautas Magnus University, Faculty of Informatics, Kaunas, Lithuania.

Neuman, J. and Morgenstern, O. (1953). *Theory of Games and Economic Behavior.* Princeton University Press, Princeton.

Owen, G. (1968). *Game Theory.* W.B. Saunders, Philadelphia, PA.

Perlibakas, V. (1999). Software for the walras problem. Technical report, Kaunas Technological University, Faculty of Informatics, Kaunas, Lithuania. (in Lithuanian).

Petrikis, S. and Juodgalvyte, J. (1997). Java1.1 software for the portfolio problem. Technical report, Vytautas Nagnus University, Faculty of Informatics, Lithuania, Kaunas. (in Lithuanian).

Petrosyan, L. (1985). *Cooperative Dfferential Games.* Tomsk University Publishing, Tomsk, Russia. (in Russian).

Petrosyan, L. (1993). *Dfferential Games of Pursuit.* Worls Scientific Publishing Co., River Edge, NJ.

Rao, T. S. and Gabr, M. (1984). An introduction to bispectral analysis and bilinear time series models. In *Lecture Notes in Statistics*, number 24. Springer- Verlag, Berlin.

Rastrigin, L. (1968). *Statistical Methods of Search.* Nauka, Moscow. (in Russian).

Raudys, A. and Mockus, J. (1999). Comparison of arma and multilayer perceptron based methods for economic time series. *Informatica*, 10(2):231–243.

Rodin, E. Y. (1987). A pursuit-evasion bibliography. *International Journal of Computers and Mathematics with Applications*, 13:275–340.

Rosenmuller, J. (1981). *The Theory of Games and Markets.* North-Holand, Amsterdam.

Saltenis, V. (1971). On a method of multi-extremal optimization. *Automatics and Computers (Avtomatika i Vychislitelnayya Tekchnika)*, (3):33–38. (in Russian).

Saltenis, V. (1989). *Analysis of Structure of Multi-modal problems.* Mokslas, Vilnius, Lithuania. (in Russian).

Schittkowski, K. (1985). A fortran subroutine solving constrained nonlinear programming problems. *Annals of Operations Research*, 5:485–500.

Senkiene, E. (1980). Properties of conditional mean and variance of the wiener process in the presence of noise, and the convergence of bayesian optimization algorithms. In *Theory of Optimal Decision*. Institute of Mathematics and Cybernetics, Akademia Nauk Lithuanian SSR, Vilnius, Lithuania. (in Russian).

Sobolj, I. (1967). On a systematic search in a hypercube. *SIAM Journal on Numerical Analysis*, 16:790–793.

Sowel, F. (1992). Maximum likelihood estimation of stationary univariate fractionally integrated models. *Journal of Econometrics*, 53:165–188.

Tiesis, V. (1975). The method of variable metrics for local optimization of functions of many variables with rectangular constraints. In *Proceedings of the Conference on Computers*, pages 111–114, Kaunas, Lithuania. (in Russian).

Tijms, H. (1994). *Stochastic Models. An Algorithmic Approach*. Wiley, New York.

Torn, A. and Zilinskas, A. (1989). *Global optimization*. Springer-Verlag, Berlin.

Torn, A. A. (1990). Topographical global optimization. Technical report, Abo Akademi University, Dept. of Computer Science, Turku, Finland.

Venclovas, A. (1999). Applying cgi scripts to upload users data in school scheduling models implemented in java. Technical report, Vytautas Magnus University, Faculty of Informatics, Kaunas, Lithuania.

von Neumann, J. and Morgenstern, O. (1953). *Theory of Games and Economic Behaviour*. Princeton University Press, Princeton, New Jersey.

Wald, A. (1947). *Sequential Analysis*. J. Wiley, New York.

Wald, A. (1950). *Statistical Decision Functions*. J. Wiley, New York.

Walras, L. (1874). *Elements d'Economie Politique Pure*. LKorbaz and Company, Lousanne.

Yin-Wong, C. and Lai., K. (1993). Long-run purchasing power parity during the recent float. *Journal of International Economics*, 34:181–192.

Index

GPSR Compliance

The European Union's (EU) General Product Safety Regulation (GPSR) is a set of rules that requires consumer products to be safe and our obligations to ensure this.

If you have any concerns about our products, you can contact us on ProductSafety@springernature.com

In case Publisher is established outside the EU, the EU authorized representative is:

Springer Nature Customer Service Center GmbH
Europaplatz 3
69115 Heidelberg, Germany

Batch number: 10755284

Printed by Printforce, the Netherlands